Politics and Development
in a Transboundary Watershed

Joakim Öjendal • Stina Hansson • Sofie Hellberg
Editors

Politics and Development in a Transboundary Watershed

The Case of the Lower Mekong Basin

 Springer

Editors
Joakim Öjendal
University of Gothenburg
School of Global Studies
Konstepidemins väg 2A-E
413 14 Gothenburg
Sweden
joakim.ojendal@globalstudies.gu.se

Stina Hansson
University of Gothenburg
School of Global Studies
Konstepidemins väg 2A-E
413 14 Gothenburg
Sweden
stina.hansson@globalstudies.gu.se

Sofie Hellberg
University of Gothenburg
School of Global Studies
Konstepidemins väg 2A-E
413 14 Gothenburg
Sweden
sofie.hellberg@globalstudies.gu.se

ISBN 978-94-007-0475-6   e-ISBN 978-94-007-0476-3
DOI 10.1007/978-94-007-0476-3
Springer Dordrecht Heidelberg London New York

Library of Congress Control Number: 2011941753

Printed on acid-free paper

Springer is part of Springer Science+Business Media (www.springer.com)

Preface

The deep roots of this book project are to be traced to a project named LMID pursuing second-track diplomacy among the Lancang-Mekong countries, promoted by SIDA/SENSA in Bangkok. Within this project a workshop was held in Göteborg by the autumn of 2008, attended by a number of key policy makers and researchers. This was the event where discussions on this topic was initiated and where the volume started its journey; a journey that turned out to be longer than anticipated, but no less rewarding then envisaged. In all, this has been an organically interesting and intellectually stimulating process.

What was painfully obvious in that initial meeting – harbouring people with a preference for analysis of regional politics as well as those with a primary interest in development – was that the discourse on participatory development of common water resources lived one life, where politicised national interest in the transboundary basin lived another: power politics vs participation, and crude economic interests vs sustainability. It seems like these are elements that do not go well together. Yet by necessity they coincide in the lived world, producing a magnificent paradox in which none seems to yield. In fact, in our observations, the trend is that the Integrated Water Resource Management approach is being deepened and that the regional rivalry over water tends to increase, seemingly sharpening the paradox that first caught our attention.

Our geographical field of interest is the Lancang-Mekong basin; a basin consisting of six countries sharing the eighth largest river in the world, all of which have major development ambitions and looking at the Mekong waters to fulfil various scenarios. Moreover, the mainstream dam-building scenario has been dormant for four decades, but is now revived, fuelled by economic growth, energy hunger and need of export incomes. This seems to us to be a key question for regional development efforts, and one that will not go away any time soon. Since the transboundary dimension of water management is increasingly coming to the fore in the global water crisis, and the IWRM is mounting into a paradigm of its own, these questions provide a generic dilemma, stretching far beyond the Lancang-Mekong region.

As mentioned, this volume has been a process in which many people and organisations have been involved. First and foremost we thank the sponsor of the workshop,

Swedish Sida and its environmental secretariat in Asia (SENSA). From SENSA, we have had the pleasure to work with Christer Holtsberg and Karin Isaksson which are our 'heroes', always trying to 'do the right thing'. Bent Jørgensen has indirectly contributed more than he knows. Hanna Leonardsson has very professionally been preparing the manuscript for publication and kept the rest of us in order. Finally, a series of unnamed reviewers have thoroughly been scrutinizing previous versions of the various components in this work. We are grateful to all. The remaining mistakes we, the editors, take responsibility for.

Göteborg   Joakim Öjendal
                                                                                        Stina Hansson
                                                                                        Sofie Hellberg

Contents

Contributors

David J.H. Blake School of International Development, Faculty of Social Sciences, University of East Anglia, NR4 7TJ Norwich, UK, djhblake@yahoo.co.uk

Rachel Cooper Newcastle University, Newcastle Upon Tyne, Tyne and Wear NE1 7RU, UK, rachel.cooper@newcastle.ac.uk

John Dore Senior Water Resources Advisor – Mekong Region, Australian Agency for International Development (AusAID), Vientiane, Lao PDR, johndore@loxinfo.co.th

David S. Hall Mekong Sub-region Social Research Centre, Ubon Ratchatani University, Ubon Ratchathani, Thailand

Stina Hansson School of Global Studies, University of Gothenburg, Box 700, 405 30 Gothenburg, Sweden, Stina.hansson@globalstudies.gu.se

Sofie Hellberg School of Global Studies, University of Gothenburg, Box 700, 405 30 Gothenburg, Sweden, Sofie.hellberg@globalstudies.gu.se

Philip Hirsch School of Geosciences, The University of Sydney, Madsen Building (F09), Rm 472, Sydney, NSW 2006, Australia, philip.hirsch@sydney.edu.au

Kurt Mørck Jensen Danish Institute for International Studies (DIIS), Strandgade 56, Copenhagen, Denmark, kumoje@gmail.com

Mira Käkönen Water & Development Research Group, Aalto University, P.O. Box 15200, FIN-00076 Aalto, Finland, mira.kakonen@utu.fi

Marko Keskinen Water & Development Research Group, Aalto University, P.O. Box 15200, FIN-00076 Aalto, Finland, marko.keskinen@aalto.fi

Matti Kummu Water & Development Research Group, Aalto University, P.O. Box 15200, FIN-00076 Aalto, Finland, matti.kummu@aalto.fi

Kate Lazarus Water Governance Specialist and Mekong Coordinator, M-POWER and the Challenge Program on Water and Food, Vientiane, Lao PDR, katelazarus2008@gmail.com

Darrin Magee Environmental Studies, Hobart and William Smith Colleges, Geneva, NY 14456, USA, magee@hws.edu

Naho Mirumachi Department of Geography, King's College London, Strand, London WC2R 2LS, UK, Naho.mirumachi@kcl.ac.uk

Joakim Öjendal School of Global Studies, University of Gothenburg, Box 700, 405 30 Gothenburg, Sweden, Joakim.ojendal@globalstudies.gu.se

Worawan Sukraroek School of Geosciences, University of Sydney, Madsen Building (F09), NSW 2006, Australia, s_warawan@yahoo.com

Ashok Swain Department of Peace and Conflict Research, Uppsala University, Post Box 514, SE-751 20 Uppsala, Sweden, Ashok.Swain@pcr.uu.se

Olli Varis Water & Development Research Group, Aalto University, P.O. Box 15200, FIN-00076 Aalto, Finland, olli.varis@aalto.fi

Abbreviations

AC	Alternating Current
ADB	Asian Development Bank
BDP	Basin Development Plan
BDP2	2nd phase of the Basin Development Programme
CIA	Cumulative Impact Assessment
CPWF	Challenge Program on Water and Food
DRIFT	Downstream Response to Imposed Flow Transition
DSF	Decision Support Framework
E-flows	Environmental flows
EFA	E-flows assessment
EGAT	Electricity Generating Authority of Thailand
EP	Environment Programme
FWRM	Fragmented Water Resources Management
FYP	Five-Year Plan
GEF	Global Environment Fund
GMS	Greater Mekong Subregion
GWP	Global Water Partnership
HRPMB	Huong River Projects Management Board
IBFM	Integrated Basin Flow Management
ICCON	International Consortium for Cooperation on the Nile
IKMP	Information and Knowledge Management Programme
IMC	Interim Mekong Committee
IPCC	Intergovernmental Panel on Climate Change
ISH	Initiative on Sustainable Hydropower
IUCN	International Union for Conservation of Nature
IWMI	International Water Management Institute
IWRM	Integrated Water Resource Management
JC	Joint Committee
JPoI	Johannesburg plan for implementation
KCM	Kong-Chi-Mun

LMB	Lower Mekong Basin
LRGR	Longitudinal Range Gorge Region
LSRB	Lower Nam Songkhram River Basin
MC	Mekong Committee
MEP	Ministry of Electric Power
MRC	Mekong River Commission
MRC/WUP	Mekong River Commission Water Utilisation Programme
MRCS	MRC Secretariat
MSP	Multi-stakeholder platform
NARBO	Network of Asian River Basin Organisations
NBI	Nile Basin Initiative
NDRC	National Development and Reform Commission
NGPES	National Growth and Poverty Eradication Strategy
Nile-COM	Council of Ministers of Water Affairs of the Nile Basin
Nile-SEC	Nile Basin Secretariat
Nile-TAC	Nile Basin Technical Advisory Committee
NMC	National Mekong Committee
PNPCA	Procedures for Notification, Prior Consultation and Agreement
RBO	River Basin Organisations
SADC	Southern African Development Community
SEA	Strategic Environmental Assessment
SEARIN	Southeast Asia Rivers Network
SIMVA	Social Impact Monitoring and Vulnerability Assessment
SOE	State-Owned Enterprise
SPCC	State Power Corporation of China
StM	Save the Mekong
TE	Trapping Efficiency
TERRA	Towards Ecological Recovery and Regional Alliance
TNC	The Nature Conservancy
UHV-DC	Ultra-High-Voltage Direct Current
UNDP	United Nations Development Programme
UNECAFE	United Nations Economic Commission for Asia and the Far East
WB	World Bank
WCD	World Commission on Dams
WUP	Water Utilization Project
ZACPLAN	Zambezi Action Plan
ZRA	Zambezi River Authority

Chapter 1
Politics and Development in a Transboundary Watershed: The Case of the Lower Mekong Basin

Stina Hansson, Sofie Hellberg, and Joakim Öjendal

Abstract This introductory chapter presents the main theme of the volume: the perceived dilemmas in pursuing IWRM in a transboundary context. The chapter discusses the IWRM approach and its package of progressive values and practices that focus on integration and participation and contrast it to transboundary politics and its tendency to remain within a state logic that emphasises sovereignty and national interests. In order to realise sustainable, efficient and inclusive water management, the chapter argues that it is essential to recognise and visualise power asymmetries and politics in regional water politics. Based on this assumption – that politics matter – the chapter contends that there is a need to explore how the perceived dichotomy between the interests of state sovereignty and (progressive) transboundary water management is played out in the Mekong River Basin. Together with its 50-year history of institutionalised cooperation and the river basin's significance in terms of supporting local livelihoods and its contribution to the region's national economies, the case is of paramount importance and interest. The disputed results and uncertain future in the region illustrate the complexity of achieving efficient, equitable and ecologically sustainable water management in a competitive international system. It thus makes up an excellent case study to illuminate the politics of IWRM in a transboundary setting. The different chapters of the volume, which are set to unpack, scrutinise, and illuminate the politics of the Lancang-Mekong Basin, are introduced at the end of the chapter. This section thus indicates some of the possible ways forward, challenges, dilemmas and incompatibilities in sustainable water management in the region which will be dealt with in more depth throughout the book.

S. Hansson (✉) • S. Hellberg • J. Öjendal
School of Global Studies, University of Gothenburg, Box 700, 405 30 Gothenburg, Sweden
e-mail: Stina.hansson@globalstudies.gu.se; Sofie.hellberg@globalstudies.gu.se;
Joakim.ojendal@globalstudies.gu.se

J. Öjendal et al. (eds.), *Politics and Development in a Transboundary Watershed:*
The Case of the Lower Mekong Basin, DOI 10.1007/978-94-007-0476-3_1,
© Springer Science+Business Media B.V. 2012

1.1   Introduction

In March 2010, large parts of the Mekong River Basin faced one of the worst droughts in decades. Agriculture, food security, access to clean water and river transport were all affected, and this threatened local livelihoods as well as national economies (MRC 2010). Farmers in northern Thailand have reportedly been fighting over water supplies, and administrators have called for negotiations with China to release more water from the dams in Yunnan province (Bangkok Post 4/3 2010). The effects of changing water patterns, where climatic variations are aggravated and complicated by large-scale dam development and other economic activities, highlight not only the urgency but also the difficulty of transboundary cooperation and integrated solutions in a complex ecological and political system such as the Mekong Basin. An abundance of water and what the MRC (Mekong River Commission), the World Bank and the Asian Development Bank (ADB) describe as the considerable developmental space of the Mekong have allowed business based on national interest and the imperative of economic growth to continue as usual (cf. Hang and Ton 2008). Yet, the increasingly strained issue of water access brings the political aspects of prioritisations and trade-offs to the fore. Moreover, the role and responsibility of the MRC as a 'mediating' institution in transboundary cooperation also becomes accentuated and possibly politicised.

Water and its governance are crucial topics in the Mekong region and globally. It is widely recognised that when water becomes commodified and supply becomes limited due to population growth, agricultural expansion, pollution, urbanisation and industrialisation, better mechanisms for managing, developing and delivering water are urgently needed. Water is an essential resource, and access to it impacts upon economic growth, poverty reduction and livelihoods. The fact that on global level, the lion's share of accessible freshwater supplies is found in transboundary systems makes the task of water management all the more challenging. However, the knowledge generated about how to deal with transboundary waters is fraught with weaknesses, and there are few successful cases (cf. Earle et al. 2010).

The dominant approach to the development of water resources is the *IWRM* approach ('Integrated Water Resource Management' – GWP 2000; cf. Conca 2006: Chap. 5; Medema and Jeffrey 2005). This consists of a package of progressive values and practices that focus on integration and participation and enable well-considered development of available water resources in any system (see below). By contrast, international/regional politics may adhere to a state logic that emphasises sovereignty and national interests instead (Hirsch and Jensen 2006). These two different views seem to be difficult to combine, but in a transboundary basin, they inevitably come to the fore. Although they seemingly contradict each other, their co-existence takes many forms and produces uncertain outcomes that need to be empirically scrutinised in each case.

This co-existence may reinforce state-based economic growth models and national boundaries in water management. However, if it makes politics visible, it may also have the potential to enable the progressive development of water resources.

This in turn may provide an opportunity for constructive debate on the future of water resource use and management. The major purpose of this volume is to illuminate shades and variations of this dilemma. The case of the six-country Mekong Basin, with its half a century history of institutionalised cooperation, disputed results and uncertain future, illustrates the complexity of achieving efficient, equitable and ecologically sustainable water management in a competitive international system.

While the duality between an IWRM approach and transboundary politics needs to be unpacked and scrutinised, the bottom line of transboundary water management is that we are still waiting to see an approach or case that has sustainably bridged these two fundamentally different perspectives. Nevertheless, policy processes and strategic decisions continue to be made as though the combination was trivial and a positive outcome assured. Or as Edkins et al. (1999) would phrase it, IWRM is 'narrated as always already inevitable' (taken from Mannergren-Selimovic 2010:8).

This volume aims to disentangle this predicament by probing the dilemma from various angles and providing new perspectives and empirical insights. Below, we will unpack IWRM by paying attention to embedded dimensions of power and politics. Then we invert our perspective and ask how 'development' may be viewed in a transboundary context. This introductory chapter then zooms in on transboundary governance and transboundary governance in the Mekong region, respectively. The chapter is concluded with a review of the various contributions to the volume.

1.2 The Politics of IWRM

The IWRM approach is the antithesis of conventional, fractional and fragmented water management systems. Emphasis is put upon *integration* and *coordination*. In the oft-quoted definition by the Global Water Partnership (GWP) in its seminal 2000 publication, IWRM is defined as

> ....a process which promotes the co-ordinated development and management of water, land and related resources in order to maximise the resultant economic and social welfare in an equitable manner without compromising the sustainability of vital ecosystems. (GWP-TAC 2000[1])

Equity, *efficiency* and *environmental sustainability* thus make up 'the essence of IWRM' (Swatuk 2005). In water management, 'business as usual' is therefore 'neither environmentally sustainable, nor is it sustainable in financial and social terms' from an IWRM perspective (GWP homepage 2010). This approach takes social, economic and environmental issues into account, thus making these three dimensions, or the three E's, decisive.

---

[1] This is elegantly picked apart by Biswas, arguing that the definition is simply unusable (Biswas 2008a:9). Yet, it is used rhetorically as well as in practice in a myriad of ways depending on preferences and interests. Jeffrey and Gearey (2006) provide ample examples of Biswas' point.

GWP has assumed a leading advocacy role in the spreading of IWRM principles, and it followed up its 2000 publication by operationalising it; the three pillars of IWRM were summarised as (1) moving towards an *enabling environment* of appropriate policies, strategies and legislation for sustainable water resources development and management; (2) putting in place the *institutional framework* through which the policies, strategies and legislation can be implemented and (3) setting up the *management instruments* required by these institutions to do their job (GWP-TAC 2004, from Medema and Jeffrey 2005).

The IWRM approach has been immensely influential and is typically seen as the way forward. Jeffrey and Gearey claim: 'It is difficult to overstate the extent to which IWRM has become the norm or even, one might say, the orthodoxy in water resources management' (2006:2). The 2005 Johannesburg Plan of Implementation (JpoI) (JpoI 2005) called for all countries to develop integrated water resources management and water efficiency plans by 2005. With support to developing countries and national/regional strategies, plans and programmes should be developed and implemented 'with regard to integrated river basin, watershed and groundwater management' (JpoI 2005:15). Progress in water management, as formulated in global policy documents, is thus measured in relation to the degree of compliance with the stipulations of IWRM (see e.g. World Bank 2004; UN WWDR 2003; UN WWDR-2 2006). It is thus no exaggeration to state that the term is a global call for integration in water management.

However, there is a gap between the ideals of IWRM and the degree of implementation. The complexity in constructing water management systems according to ingredients of IWRM, the trade-offs involved, as well as the limitations of the approach at the level of implementation have been widely discussed and critically assessed (e.g. Biswas 2008a, b; Jeffrey and Gearey 2006). Critics of the IWRM approach have argued that given the complexities in the world's river basins, the 'one-shot approach of management within the context of IWRM is far too simplistic to be useful, or applicable' (Varis et al. 2008: xii). They have also pointed out that there is a lack of acknowledgement of the antagonistic relationship between the three E's of IWRM and that the trade-offs between them will be results of difficult political processes that involve many actors, institutions and objectives, and they are often characterised by unequal power relations. These, in turn, become obstacles to the achievement of optimal allocation of water resources, and they do not provide satisfactory political and social solutions (Molle 2008: 133; cf. Zeitoun and Warner 2006).

Contrary to the intentions of IWRM, it has also been argued that the approach allows for 'business as usual' since the broad and inclusive character of IWRM can be used to promote interests ranging from private interests to the interests of livelihood-oriented NGO and social or environmental activists (Biswas 2008a; Molle 2009: 134). IWRM may also buy into donors' preferences and the entire political economy of aid and state building, which are distant from water management as such. It may therefore be argued that IWRM does not question but instead reinforces the traditional roles, mandates and worldviews of central actors in water management (Molle 2009: 135). It transforms into a 'cover-up' under which any agenda and interest can be pursued.

In fact, critical perspectives abound. For instance, research in Southern Africa has shown that rather than solving problems of increased competition over water resources domestically, IWRM-inspired policies have in fact worsened inequality in access to water and have favoured industry and large-scale farming over subsistence agriculture and the domestic needs of poor households (see e.g. McDonald and Ruiters 2005; McDonald and Pape 2002; Bond 2000; Loftus 2005a; 2005b; Merrey and van Koppen 2007). Despite uncertainty about the success of the implementation of the IWRM approach, donors continue to pump money into IWRM-related activities, apparently confident that positive results will be achieved (Biswas 2008b). IWRM seems to represent the technocratic dream of the 'machine', nullifying politics and wishing away power and interests. Of course, this will not happen either for water governance or for the development process at large, as critics have pointed out (cf. Scott 1998; Molle 2009).

However, IWRM is something of a 'moving target'; new problems emerge and evolve over time (cf. Molle 2008), so a flexible approach could allow for learning from past experiences. Similarly, on the GWP homepage, it is written that 'IWRM has no fixed beginnings and will probably never end' (GWP homepage 2010). This would mean that we are in a situation of constantly (re)evaluating, (re)creating and perhaps (re)negotiating water management systems. This enables 'adaptive management' that has more feedback loops and allows for long-term uncertainties and perhaps greater need of acknowledgement of the presence of politics and interests (Medema and Jeffrey 2005).

However, viewing IWRM from this perspective demands that we abandon the concept as a technical and managerial tool and that we recognise its political dimensions. This also begs the question of who should be 'doing' IWRM – national water agencies, river basin organisations or technical experts? Or should it be a popularly driven process that is directed into the general IWRM framework? Furthermore, how will any IWRM process be steered without initiating the re-centralising processes that were to be avoided in the first place? Hence, not only is IWRM paradoxical, so is its process (Biswas 2008b).

Moreover, on top of the complexity of the politics of IWRM in states, transboundary contexts may also become arenas of struggle about national interests; if, on the one hand, IWRM aims to 'hide' the politics of water management, on the other, transboundary water management is defined by its vested interests. Typically, the dominant actors in the system have low incentives and few mechanisms for abandoning these interests. Having now introduced politics, interests and power into the perception of IWRM, let us turn to the way in which development might be understood in a transboundary context.

1.3 Development in a Transboundary Context

Whether we look at the Nile, the Ganges, the Danube, the Euphrates–Tigris, the Amu Darya or Orange basins, the transboundary nature of these basins provides a challenge to efficient and sustainable development processes that enable their full potentials to be tapped, gains shared, ecological systems preserved and disputes managed (cf. Swain 2004; Huitema and Becker 2005; Varis et al. 2008).

Because of its multifaceted and crucial short- and long-term economic, social and political importance, water management is difficult even in domestic contexts. In transboundary settings, issues of politics, power and security are still more problematic (Earle et al. 2010). This means that water management is liable to become securitised (Turton 2001; Turton and Funke 2008) and thus distanced from sustainable, inclusive, joint and/or participatory management. Power and politics are present also in domestic settings, particularly in multi-ethnic states and in states with an advanced federal system. However, the 'sovereignty' dimension is what makes transboundary settings exceptionally complex and prone to securitisation and conflict.

In a narrow sense, the process of securitisation may be understood to increase the risk of conflict, as is captured in the discourse on water wars (cf. Starr 1991). However, we do not believe that all transboundary basins are bound to be the sources of military conflict, violence or misery. On the contrary, a modern mantra in the field of water management is that cooperation is the dominant pattern among states surrounding these basins (Wolf et al. 2003; cf. Phillips et al. 2006; cf. Swain 2004). While this mantra is important, it is not entirely convincing. 'Cooperation' is a fungible term (Oye 1985; Axelrod 1984; Axelrod and Keohane 1985; Jägerskog 2003; Swain 2004; Phillips et al 2006; Earle et al. 2010; Mirumachi 2010), and the pattern seen in Wolf et al.'s historical database obscures several facts; although 'war' seldom results from water rivalry, lower-intensity conflicts often do (cf. Wolf et al. 2003; Pryor 2007), and in most of this literature, cooperation is simply regarded as the signing of agreements or rhetoric that may obscure rather than demonstrate genuine commitment. Cooperation of this kind is often shallow and may be used to justify doing nothing; no joint development efforts follow, and the contents of the agreement are only loosely respected (see Swain, this volume). The dominant pattern is that most water agreements are not followed by implementation of that which has been agreed. Instead, as has been shown in much of the literature and in these databases, this is really conflict avoidance (cf. Phillips et al. 2006).

Securitisation may be understood more broadly from a post-structuralist IR perspective, and this brings other effects to the fore. When water scarcity is constructed as a threat to the economic development of the state, it may be made into an issue of national security. This may then provoke conflicts between states or between interest groups within states. Defining the management of the water resource as a matter of the state's survival allows the state to grant itself special rights in governing it (Wæver 1995; Cascao and Mark 2010; cf. Turton and Funke 2008). Water becomes caught up in a logic of national interest that goes well beyond the resource as such. This logic includes everything from energy production to financial stability and state 'survival' (Biswas 2008a). The perceived imperative of enhancing water access and channelling it towards economic development in order to strengthen the state legitimises the state's overruling of social and ecological considerations or other legitimate political interests. This means that those who advocate integrated and sustainable management become excluded from decision making, and this has consequences for political legitimacy and poverty alleviation. The embeddedness of this rationality in international relations and an international economic system based on competition yields a logic that needs to be unpacked to enable constructive solutions.

The state development logic, which is based on large-scale economic growth and modernisation, has consequences for minority groups and small-scale livelihood sustainability. Alternative perspectives and interests may play only minor roles in the state-wide development perspective, or they may even be constructed as security threats that impede modernisation and growth. As such, they may be targeted for state interventions to exclude them or integrate them into the modernising state (Scott 1998). Even within governments, various positions may compete for influence, and this may determine the outcome of efforts to cooperate and integrate.

Cooperation, in the form of transboundary agreements and river basin organisations (RBOs), may reduce the risk of (violent) conflict but remain within a water management logic that is based on sovereignty and national interest, and this may legitimise rivalry over resources. There is clear evidence that states are unwilling to yield part of their sovereign responsibility for water resources to international bodies (Turton and Funke 2008:10; Australian Mekong Resource Center AMRC and Danida 2006). Even when transboundary cooperation exists, individual states may act together to promote the same logic; they may agree that national interests in economic growth may best be served through joint infrastructure projects and so on. This kind of regional consensus between states may further marginalise other voices and interests as transboundary megaprojects are implemented, benefitting the urban upper and middle classes while the needs of local groups are ignored.

This hegemonic discourse has been further reinforced by the fact that transboundary agreements and river basin organisations render water management a technical rather than a political issue. Doing this has been necessary in order to enable agreements to be reached, and it also tends to de-securitise water and make the area less conflict prone (Turton and Funke 2008). However, in reality, instead of resolving rivalry between states, this process may simply conceal contentious issues from view and make it possible to continue 'business as usual' under the smoke screen of consensus (Molle 2008). Viewed from a structural angle, the net effect of the processes described above is that there is typically politicisation and/ or securitisation on the national level while there is de-politicisation at the regional level, and this creates a mismatch and miscommunication between individual states and bodies working at the regional level. Moreover, donors who set up agreements and RBOs contribute to de-politicisation by trying to avoid the quagmire of politics. They thereby run the risk of feeding into pre-existing logics and power relations. Donors have pursued the IWRM agenda enthusiastically, but it has been argued that they nevertheless perpetuate the state logic because they rely on the state for decision making and implementation and they regard it as holding ultimate responsibility for water management. According to Molle, this means there is a 'high likelihood of reproducing paternalistic, technocratic and bureaucratic top-down conventional approaches, modified only by whatever degree of participation is allowed' (Molle 2008:134).

The integrative and participatory development ambitions of IWRM are thus at risk of becoming marginalised as other logics come to dominate; at the same time, its de-politicising effects obscure the way in which 'business as usual' continues with little regard for IWRM ideas.

1.4 Transboundary Water Governance

It is ultimately the political choices of riparian countries that will decide the fate of the river. (Australian Mekong Resource Centre and Danida 2006)

'Shallow cooperation', understood as conflict avoidance and the tendency to do 'nothing' after signing agreements, may impede collaborative sustainable development efforts in transboundary water governance. Furthermore, as water becomes increasingly scarce worldwide, intensifying securitisation may hinder progressive ways of utilising water resources (cf. Phillips et al. 2006). This persists because the state-focused perspective tends to limit thinking beyond the imperatives of macro-economic growth and inter-state rivalry over water, and this, in turn, limits the development of strategies that are designed to achieve more equitable and sustainable use of water.

However, adopting a state-centred perspective on transboundary waters may lead to a belief that IWRM and transboundary basins are incompatible; this belief needs to be scrutinised. There is no such thing as a single national interest – there are always competing national interests. The power relations between interest groups within states must be explored in any discussion of sustainable management of the river. Although some IR scholars have noted that 'politics matter' in transboundary water management (Lowi 1993; Allan 2001; Warner and Zeitoun 2008), in general, politics remain hidden, forgotten or neglected (Furlong 2006). Warner and Zeitoun note that '… the number of serious studies applying IR frameworks to transboundary water issues remains limited'. In addition, we have the persistent gap in how rationally the donor community understands the post-colonial state, contrasted to their perceived political needs and its close relation to their national identity and survival.

While comprehensive IR studies are few, there are fewer still that take a critical perspective (Furlong 2006:803). The absence of this does not indicate that the problems emanating from international dynamics do not exist, only that they are neither commonly nor properly addressed. Jeffrey and Gearey, approaching the issue from the IWRM perspective, explain how the lack of attention to political matters accounts for the relative failure of a widespread implementation of IWRM, and Wostl et al. claim that in transboundary management, '… there persist major gaps between scientific and political rhetoric and the implementation of change at the operational level' (2006:4). Kranz et al. add:

… other findings indicate that in transboundary regimes the discussion of the 'politics of (water management) policy', i.e. the overall hydro-political dimension of resources management, is of high relevance.......The comparison between countries conducted [here] shows that differences in socio-economic situation (and the (military) power connected to them), political situation (allowing for participation or not) and geographical situation (upstream – downstream) have a direct impact on water management policies, and that the knowledge thereof can be conducive to developing (good) co-operation. (Kranz et al. 2005: 16)

This understudied quality of the development of transboundary water resources is at the core of this volume and will be illuminated in each of the chapters.

The other problem in this dilemma of managing transboundary water resources is the establishment of institutions or regimes (Huitema and Becker 2005; Young 1989). These are expected to accommodate conflicting interests, provide quasi-diplomatic channels and craft technically sound and economically agreeable solutions. 'Institutions' may include political traits, international laws, water agreements and/or river basin organisations (RBOs) (Huitema and Becker 2005). The Tennessee Valley Authority (TVA) is a paradigmatic example that has been exported to and followed in the global South for at least half a century (Öjendal 2000). However, the success of these institutions can be described as mixed, at best. With growing water scarcity, global climate change (Falkenmark and Jägerskog 2010; Drieschova et al. 2009), growing demands of economic growth, widespread public protests over water policies and calls for increased participation as well as for efficient project management (Conca 2006), these hardly provide a comprehensive solution to the problems encountered in transboundary water management. In fact, no single package can provide a final solution since the availability, value and importance of water vary geographically, politically and culturally. It may therefore be better to acknowledge these differences and the need for flexibility.

So while there are positive trends to build on, the challenges of cooperation and its institutions need to be better understood in order to facilitate a progressive use of water resources in a context that is characterised by a variety of interests and of unequal power relations between stakeholders. There is a risk that rather than applying IWRM principles so that integration, participation and efficiency result, the policies and practices of these institutions may be caught up in the logic of sovereign state interests. In this way, they may unwittingly exacerbate the incompatibility between IWRM and transboundary water management.

Focusing on the contradiction between IWRM and transboundary water management may hide dilemmas of water management *within* states, dilemmas that may resemble each other in the different countries. One example is the sustainability of small-scale farming in relation to large-scale infrastructure and investments; the competing interests in fishing and in dam construction are another. As a progressive tool, IWRM would require not only 'cooperation' between states but also integration and participation by small-scale farmers, civil society and environmental advocates both within and between states.

Based on the assumption that politics matter, we would like to explore how this perceived dichotomy between the interests of state sovereignty and (progressive) transboundary water management is played out in the Mekong River Basin. While a critical approach is called for (cf. Allan 2003; Molle; 2008), we would like to also reflect on whether the incompatibility described above may be overcome, how it may be worked with and how workable compromises might emerge. This is something that research just recently turned attention to (cf. Earle et al. 2010). Our case is illuminating the dilemmas of the development of sustainable transboundary water management and how the intense attempts to solve them can be construed (see below).

1.5   The Mekong Basin as a Transboundary System

Our case – the Mekong River Basin[2] – is of paramount importance, and it well illustrates the problem. The Lancang-Mekong is situated in southwestern China and mainland Southeast Asia. It is one of the major rivers in the world, ranking number eight in terms of flow, covering an international basin of six countries[3] and 795,000 km[2] that are inhabited by some 80 million people. The importance of the Mekong Basin is not the water alone but the ecosystem in full as it sustains some 80% of the basin's inhabitants (Fox and Sneddon 2005:2). The majority of the population live in rural areas sustaining themselves on small-scale primary production.[4] Large-scale agro business in Thailand and the Vietnam Delta constitute the main exceptions. Against this background, cooperation and planning of the use of Mekong water resources become imperative.

The Mekong Basin is one of the poorest areas in the world, where a quarter of the population is estimated to be living below the poverty line (Kaosa-ard 2003:84), and the GNI per capita ranged between US $480 in Cambodia and US $2,990 in Thailand in 2006 (Varis et al. 2008). The development imperative is urgent, and the Mekong River Basin has become a focus area for economic growth and development in mainland Southeast Asia over the past decade.

The water resources of the Mekong have not yet been harnessed by modernisation projects, though there are major interest lining up to alter this, e.g., the Yunnan

---

[2] The 'Mekong' is, with local variations, the accepted name of the river in the Lower Basin (Thailand, Laos, Cambodia and Vietnam); in the upper reaches – i.e. China – it is called the Lancang River. We have our focus in this volume on the lower basin cooperation/development, and we have therefore adopted the 'Mekong' name for the river.

[3] The river originates in Tibet, China, and flows for a long stretch through China where it also falls rapidly. It touches upon Myanmar, before it enters Laos, constituting a border river to Thailand for a while, returning into Laos and then entering Cambodia. In Cambodia, the rapid flow slows as it enters the plains before reaching the delta, southeast of Phnom Penh. The major share of the delta is in the southernmost part of Vietnam. The upper part of the river, situated in China, is called Lancang; the full name of the river basin is then the Lancang-Mekong River Basin.

[4] While this is correct, beneath the statistics, there is a wealth of other livelihoods, including migrant workers who engage in other economic activities, which means that many households and communities are also dependent on incomes generated in other parts of the economy.

Cascade[5] (cf. Keskinen; Magee this volume); the Lower Mekong riparians also have a large number of projects in the pipeline. About ten large hydropower projects are under construction, and almost one hundred fifty projects are being planned, including eleven on the mainstream (Hang and Ton 2008:2). The area is in need of major development, but it is also sensitive to change; local livelihood systems are often based on the existing ecological system. Development projects in all riparian countries – often pursued in a crude modernising fashion and with little regard for local participation or preferences – pose various threats to the ecological system which is further aggravated by global warming.

The WB/ADB joint working paper on Future Directions for Water Resources Management in the Mekong River Basin (June 2006) states that

> … there remains considerable potential for development of the Mekong water resources. The Mekong basin has flexibility and tolerance, which suggests that sustainable, integrated management and development can lead to wide-spread benefits.

This, they argue, contrasts with the 'more precautionary approach of the past decade' (2006:4). The conclusion is that 'balanced development' should be the driving principle – implying trade-offs 'between economic, social and environmental values; between the competing interests of the riparian countries; and between the different sectors and beneficiary groups at the sub-basin level'. They emphasise the need to seek win-win situations, which, it is argued, can be achieved by applying IWRM. In such a context, the IWRM becomes a technical managerial device with de-politicising ambitions that limits the potential for integrated solutions.

The Mekong area has a history of conflicts, including a number of the most violent ones since the Second World War, and these did not come to a halt until peace was established in Cambodia in 1993 (in reality, in 1998).[6] Hence, regional relations, state building and inter-state cooperation are anything but simple. The two aspects we are concentrating on here – development efforts in the light of transboundary complications – are present in ample measure.

Mekong River planning dates back to more than 50 years and has gone through a range of phases and political negotiations. The most acclaimed results are extensive data gathering and dissemination of information concerning the basin's ecological and physical systems as well as its (disputed) role as a platform for dialogue on common issues (Jacobs 1995). The Lower Mekong riparian countries have received international acclaim and prizes for their ambitious take on transboundary governance, in particular through the Mekong River Commission (MRC). This is

---

[5] Since the 1990s, China has built two major dams on the mainstream, and another twelve are in the pipeline. At least two of these are expected to be among the largest in the world, i.e. the Jinghong and the Xiaowan (Hang 2008). The effects of these dams are as yet unknown, but they are of major importance to downstream countries.

[6] In fact, during 2009/2011, a minor war flared between Cambodia and Thailand, reminding us of the instability of the region.

typically seen as 'successful' and a case to be learnt from for river basin management in the South. However, this view may be a little too rosy. As Keskinen et al. say:

> The Mekong River is a good example of an international river basin that involves multiple sectors and actors and thus needs integrated management. The Mekong River Commission (MRC) has partly adopted this task, but faces many constraints such as the absence of the two upstream countries. (2008:207)

In fact, as stated in the strategic plan of MRC 2006–2010, IWRM is the strategy chosen in order to implement its strategic plan; in the fourth (of four) of its overall goals, the plan states that it shall seek 'To strengthen the Integrated Water Resources Management capacity and knowledge base of the MRC bodies, NMCs, line agencies, and other stakeholders' (MRC 2006:23–24). In particular, this should be achieved through the core programme of the Basin Development Plan (BDP, cf. Öjendal and Mørck Jensen; Hirsch, this volume) being executed during this period. In spite of sincere attempts, the MRC has not yet succeeded in becoming the key actor in the region, and this supports our contention about the way in which politicised national development efforts fail to match with de-politicised regional managerial approaches. Dore and Lazarus make the point that

> MRC has too often been absent from or silent about substantial decisions being taken on water resources development in the basin. As pointed out earlier, MRC secretariat has had little involvement and usually very limited information about the hydropower development on the Mekong River mainstream in China, and on tributaries in Laos and Vietnam. (Dore and Lazarus 2009:16)

The lack of commitment by the political leadership in member riparian states has been argued to severely limit the potential role of the MRC in the basin (Hirsch and Jensen 2006). For instance, it is convincingly (and hardly surprisingly) argued that there is a natural resistance among the riparian states to give up part of their sovereignty over shared resources. This is a legitimate position for states to adopt since the 1995 Mekong Agreement calls for '… cooperation on the basis of sovereign equality and territorial integrity in the utilisation of the water resources of the Mekong Basin' (Mekong Agreement 1995 Article 4). However, this does not move us any closer to achieving better regional water governance, which is what this volume is about.

Dore and Lazarus go on to note that 'A new water governance paradigm is needed in the Mekong Region to assist societies make better choices about how to share and manage water for production of food and energy' (Dore and Lazarus 2009:1). Understanding the potential for a more efficient as well as a more participatory water governance regime includes recognising the contradictions of an IWRM approach in a transboundary context. The contributors to this volume illustrate this in various ways.

1.6   The Contributions to This Volume

This introduction is followed by eight chapters that each illuminate the key dilemma outlined here. These are then drawn together in a concluding chapter. Firstly, *Swain* presents a theoretical overview of the field in 'Politics or Development: Sharing of

International Rivers in the South', and he adds some brief macro case studies and a few analytical notes. We learn from his contribution that in the 1990s, many agreements were drawn up in the South to share the international river basins. Of particular interest in this context are the Zambezi, the Mekong, the river Jordan, the Ganges and the Nile, which all have complex cooperation/conflict dimensions and semi-functional agreements. These agreements are, moreover, under severe stress due to increasing development pressure and uncertainties due to climate change. In these cases, the riparian countries agreed to share when there was hope of further exploitation of the river. The possibility of acquiring more water has led the political leaders to opt for the agreement as it provides political gain and development promises. However, these agreements are not good enough to initiate overall development in the basin as they lack the ability and incentives to enable the best possible use of scarce water resources to meet future water needs. Swain puts the problem of how water in a politicised transboundary basin may be used for domestic development purposes into a global context.

Öjendal and Mørck Jensen establish in 'Politics and Development of the Mekong River Basin' the empirical bottom lines of the Mekong Basin and its governance, in particular the 'gap' between, on the one hand, its 'geographies' and associated development issues and, on the other, the various international agreements designed to institutionalise and negotiate 'solutions'. Despite major efforts, these parts neither meet effectively nor deliver hopes for results. Two key projects, the Water Utilization Project (WUP) and the Basin Development Plan (BDP), are critical for realising the ambitions of the agreement. These processes have been concluded and have undoubtedly delivered interesting development and political processes, although not necessarily the ones aimed for, and they are certainly not sufficient to solve the inherent dilemmas. Finally, it is also noted that any comprehensive development planning or political agreement will be complicated by the fact that upstream interventions in China have thus far not been considered in these deliberations, and this casts more doubt upon the possibility of applying a 'clinical' IWRM approach to transboundary basin management.

Cooper's chapter 'The Potential of MRC to Pursue IWRM in the Mekong: Tradeoffs and Public Participation' explores the potential of the MRC to pursue IWRM in two key areas: the identification and negotiation of development tradeoffs and public participation. The chapter discusses the nature of the key trade-off in the Lower Mekong – the development of hydropower versus fisheries – which could bring local livelihoods and economic development into conflict. This highlights the role of the MRC in negotiating between national and transboundary interest groups and in balancing social and environmental perspectives. While this chapter argues that the MRC can generate knowledge that would bring different perspectives to this debate, it also acknowledges the challenges of doing so; it argues that knowledge generation is not enough and that the MRC will have to proactively confront issues and ensure that information has an impact on decision making. The chapter explores the possibilities and limitations of involving civil society and local communities in water management in an environment in which different stakeholders have very different chances of participating in open dialogue, both between and within countries. Ultimately, the chapter argues that it is important that the MRC strengthens its civil society engagement and finds ways to engage local communities in its work.

Mirumachi's chapter – 'Domestic water policy implications on international transboundary water development: A case study of Thailand' – focuses on the political dimension of transboundary waters and on the impact of domestic water policies, particularly in the upstream–downstream relationship between Thailand and Vietnam. Thailand, Mirumachi argues, has managed to continuously engage in cooperation while at the same time maximising its capture. The argument in this chapter is based on an analysis of the pre-1995 MRC agreement negotiations; Mirumachi shows both how national interests were played out when methods for allocating water were elaborated and how Thailand's 'hydraulic mission paradigm' became more diverse with the presence of more ecocentric concerns. She also shows how Thailand used its position as an upstream hegemon productively in an institutional setting where other avenues for implementing change were open, such as the GMS and bilateral negotiations. Hydraulic control, it is further argued, may be achieved through conflict as well as through cooperation. The author therefore stresses the importance of going beyond mere classification of conflict or cooperation as self-evident categories. The chapter shows how we can better understand the prioritisation of water issues in particular national contexts by paying attention to how different paradigms are played out, in domestic and international negotiations, and how this is linked to the political economic context.

In his chapter 'IWRM as a Participatory Governance Framework for the Mekong River Basin? Hirsch asks to what extent IWRM is a tool for enhancing participation and, if so, in which framework this will be achieved. The first observation is that historically, instead of IWRM, we have frequently seen 'FWRM' – Fragmented Water Resource Management – which appears in many different guises. In the Mekong case, the RBO (MRC) has typically been criticised for being centralised and non-inclusive. Interestingly, there have simultaneously been counter forces that strive to involve stakeholders, typically NGOs and grassroots organisations, in a broader sense. Admittedly, the MRC has – at least in some of its programmes – taken a more participatory and transparent turn with some concrete implications. IWRM may be an answer but not to the dilemma of participation in transboundary basins. Ultimately, Hirsch argues that IWRM is essentially political and that it should not be positioned in the technical realm but that it should be understood from the perspective of those living and working in the basin.

In their chapter, *'Mekong at the crossroads – alternative paths of water development and impact assessment'*, Keskinen, Kummu, Käkönen and Varis discuss different water development pathways in the Mekong Basin, their potential impacts and the possibilities of assessing them. It is argued that water development and related management practices in the Mekong are at a crossroads methodologically and, more importantly, politically. Using the example of the Tonle Sap system, the authors emphasise the cumulative impact of multiple plans in the basin and the need for a broad range of assessment models and procedures to capture the oft-neglected importance of fisheries, floodplains and other common-pool resources. Existing estimates point towards remarkable potential changes due to hydropower development, but the impacts on systems as complex as floodplains and fisheries are much more difficult to assess. The social and economic importance of the latter also alerts us to the necessity of connecting physical and ecological impacts with broader social and political dimensions. The crossroads the Mekong is now at motivates the authors' call for a participatory dialogue on future development paths. This dialogue should be based on assessments of impact as well as a discussion about distribution of benefits and costs. They suggest a pause during which stakeholder dialogue is radically strengthened, and they claim that this must be based on a debate about the politics behind the seemingly technical decisions.

In 'Negotiating Flows in the Mekong', Lazarus et al. explore the concept of Environmental Flows (E-flows) as a constructive negotiation tool for the Lower Mekong River Basin. They review the different ways that the concept has been approached by the research community, policy makers and practitioners in the region. Using several examples, the chapter discusses the challenges involved in using E-flows for river management in the Mekong, including moving from E-flows theory to practice and enabling interdisciplinary approaches as well as linking hydrological and ecological aspects. We learn from this chapter that E-flows can provide a tool for involving multiple stakeholders in dialogue to determine the best possible flow regime for the Mekong Region. However, this requires sustained support and trust building between numerous actors and institutions in order to build a critical mass of expertise and understanding as concepts are internalised.

Magee's chapter 'The Dragon Upstream: China's Role in Lancang-Mekong Development' examines China's development on the upstream half of the Lancang-Mekong River. It includes perspectives on local, regional, national and international development that inform and motivate the nature and magnitude of this development. The primary goals of the chapter are to understand Chinese development priorities for the upper half of the basin, how these priorities coincide and conflict with priorities for the downstream half and what the geopolitical ramifications of Chinese development on the Lancang-Mekong might be. Magee begins by describing the physical and human geographical characteristics of the Chinese half of the Lancang-Mekong Basin. From this vantage point, he then lays out a series of issues as they are perceived in China and shows that the corresponding solution to each problem wholly or partially justifies (from the Chinese development state's perspective) the construction of major infrastructure projects in southwestern and western China, of which the Lancang hydroelectric cascade is a major component. Finally, he discusses the extent to which China's pursuit of development objectives on the Lancang has changed in the past decade and the ramifications of such change.

Together, the empirical chapters, set against the background of the introduction and the overview by Swain, capture the politics of transboundary water governance in a politicised river basin and illuminate the complex development trade-offs these carry. It becomes clear, we contend, that in order to optimise water utilisation, we must improve our communication and dare to realise – without panicking and 'securitising' – that water management is political and that it does not respond well to attempts to hide complexities in 'Nirvana concepts' such as IWRM.

References

Allan A (2001) The Middle East water question: hydropolitics and the global economy. I. B. Tauris, London/New York

Allan A (2003) IWRM/IWRAM: a new sanctioned discourse? SOAS/KCL water issues group occasional paper 50. SOAS/King's College, London

Australian Mekong Resource Center (AMRC), Danida, Hirsch P, Jensen KM, FitzGerald S, Boen B, Lyster R, Carrard N (2006) National interests and transboundary water governance in the Mekong. Draft report Australian Mekong Resource Center

Axelrod RM (1984) The evolution of cooperation. Basic Books, New York

Axelrod RM, Keohane RO (1985) Achieving cooperation under anarchy: strategies and institutions. World Polit 38(1):226–254

Bangkok Post (2010) Drought will worsen as temperature rises to 43C. http://www.bangkokpost.com/news/local/33825/drought-will-worsen-as-temperature-rises-to-43c. Accessed 10 Apr 2010

Biswas AK (2008a) Integrated water resources management: is it working? Water Resour Dev 24(1):5–22

Biswas AK (2008b) Current directions: integrated water resources management – a second look. Water Int 33(3):274–278

Bond P (2000) Cities of gold, townships of coal – essays on South African new urban crisis. African World Press, Trenton/Asmara

Cascao A, Mark Z (2010) Power, hegemony and critical hydropolitics. In: Earle A, Jägerskog A, Öjendal J (eds) Transboundary water management principles and practice. Earthscan, London

Conca K (2006) Governing water. MIT Press, Cambridge

Dore J, Lazarus K (2009) De-marginalizing the Mekong river commission. In: Molle F, Foran T, Käkönen M (eds) Contested waterscapes in the Mekong region: hydropower, livelihoods and governance. Earthscan, London, pp 357–382

Drieschova A, Giordano M, Fishlander I (2009) Climate change, international cooperation and adaptation in transboundary water management. In: Adger WN, Lorenzoni I, O'Brien K (eds) Adapting to climate change: thresholds, values, governance. Cambridge University Press, Cambridge

Earle A, Jägerskog A, Öjendal J (eds) (2010) Transboundary water management principles and practice. Earthscan, London

Edkins J, Persram N, Pin-Fat V (eds) (1999) Sovereignty and subjectivity. Lynne Rienner, Boulder/London

Falkenmark M, Jägerskog A (2010) Sustainability of transnational water agreements in the face of socio-economic and environmental change. In: Earle A, Jägerskog A, Öjendal J (eds) Transboundary water management principles and practice. Earthscan, London

Fox C, Sneddon C (2005) Flood pulses, international watercourse law, and common pool resources: a case study of the Mekong lowlands. Research paper 2005/22, expert group on development issues. United Nations University and World Institute for Development Economics Research

Furlong K (2006) Hidden theories, troubled waters: international relations, the 'Territorial Trap', and the Southern African development community's transboundary waters. Polit Geogr 25:438–458

GWP (2010) Homepage: http://www.gwp.org/. Accessed 13 Apr 2010

GWP (Global Water Partnership) (2000) TAC background paper no.4: integrated water resources management. Global Water Partnership, Denmark

Hirsch P, Jensen KM (2006) National interests and transboundary water governance in the Mekong. AMRC: University of Sydney, Sydney

Huitema D, Becker G (2005) Governance, institutions and participation. A comparative assessment of current conditions in selected countries in the Rhine, Amu Darya and Orange basins. Institute for Environmental Studies, Amsterdam

Jacobs JW (1995) Mekong committee history and lessons for river basin development. Geogr J 161(2):135–148

Jägerskog A (2003) Why states cooperate over shared water: the water negotiations in the Jordan river basin. PhD thesis, Linköping University

Jeffrey P, Gearey M (2006) Integrated water resources management: lost on the road from ambition to realisation? Water Sci Technol 53:1–8

JPoI (Johannesburg Plan of Implementation) (2005) UN Department of Economic and Social Affairs, division for sustainable development. http://www.un.org/esa/sustdev/documents/WSSD_POI_PD/English/POIToc.htm. Accessed 20 Apr 2010

Kaosa-ard M (2003) Poverty and globalisation. In social challenges for the Mekong region. In: Kaosa-ard M, Dore J (eds) Social challenges for the Mekong region. White Lotus, Bangkok

Keskinen M, Mehtonen K, Varis O (2008) Transboundary cooperation vs. internal ambitions: the role of China and Cambodia in the Mekong region. In: Pachova NI, Nakayama M, Jansky L (eds) International water security: domestic threats and opportunities. UNU Press, Tokyo

Kranz N, Interwies E, Vorwerk A (2005) Adaptive water management in transboundary contexts – common research Agenda. Institute for Environmental Studies, Amsterdam

Loftus A (2005a) 'Free Water' as commodity: the Paradoxes of Durban's water service transformation. In: McDonald DA, Ruiters G (eds) The age of commodity: water privatization in Southern Africa. Earthscan, London/Sterling

Loftus A (2005b) A political ecology of water struggles in Durban, South Africa. PhD thesis, School of Geography and the Environment, University of Oxford, Oxford

Lowi MR (1993) Water and power: the politics of a scarce resource in the Jordan river basin. Cambridge University Press, Cambridge

Mannergren-Selimovic J (2010) Remembering and forgetting in foca – narratives of truth, justice and reconciliation in a Bosnian Town. PhD thesis, School of Global Studies, Gothenburg University, Gothenburg

McDonald DA, Pape J (2002) Cost recovery and the crisis of service delivery in South Africa. Zed Press and HCRS, Cape Town/London

McDonald DA, Ruiters G (eds) (2005) The age of commodity: water privatization in Southern Africa. Earthscan, London

Medema W, Jeffrey P (2005) IWRM and adaptive management: synergy or conflict? Institute for Environmental Studies, Amsterdam

Mekong Agreement (1995) Agreement on the cooperation for the sustainable development of the Mekong river basin. Mekong River Commission, Chiang Rai

Merrey DJ, van Koppen B (2007) Balancing equity, productivity and sustainability in a water-scarce river basin: the case of the Olifants river basin in South Africa. IWMI, Comprehensive Assessment of Water Management in Agriculture, Colombo

Mirumachi N (2010) Study of conflict and cooperation in international transboundary river basins: the twins framework. PhD thesis, LSE, London

Molle F (2008) Nirvana concepts, narratives and policy models: insight from the water sector. Water Altern 1(1):23–40

Molle F (2009) Water and society: new problems faced, new skills needed. Irrig Drain 58:205–211

MRC (Mekong River Commission) (2006) MRC work programme. Mekong River Commission, Vientiane

MRC (Mekong River Commission) (2010) Drought conditions cause low Mekong water flow. Mekong river commission, Media Release. http://www.mrcmekong.org/MRC_news/press10/drought-condition26-2-10.htm. Accessed 30 Mar 2010

Öjendal J (2000) Modes of managing the Mekong. PhD thesis, Peace and Development Research, Gothenburg University, Gothenburg

Oye KA (ed) (1985) Cooperation under anarchy. Princeton University Press, Princeton

Phillips DJ, Daoudy M, Öjendal J, Turton A, McCaffrey S (2006) Trans-boundary water cooperation as a tool for conflict prevention and for broader benefit-sharing. Ministry for Foreign Affairs, Stockholm

Pryor FL (2007) Water stress and water wars. Econo Peace Secur J 2(1):18–29

Scott J (1998) Seeing like a state. Yale University, New Haven/London

Starr J (1991) Water wars. Foreign Policy 82(Spring):17–36

Swain A (2004) Managing water conflict. Routledge, London

Swatuk L (2005) Political challenges to implementing IWRM in Southern Africa. Phys Chem Earth, Parts A/B/C 30(11–16):872–880

Turton A (2001) Towards hydrosolidarity: moving from resource capture to cooperation and alliances. Keynote address at The Stockholm International Water Institute (SIWI) symposium, Water security for cities, food and environment, Stockholm

Turton A, Funke N (2008) Hydro-hegemony in the context of the Orange river basin. Water Policy 10(2):51–69

UN WWDR (2003) World water development report. UN, New York

UN WWDR (2006) World water development report. UN, New York

Varis O, Rahaman M, Stucki V (2008) The rocky road from integrated plans to implementation: lessons learned from the Mekong and Senegal river basins. Int J Water Resour Dev 24(1):103–121

Wæver O (1995) Securitization and desecuritization. In: Lipschutz RD (ed) On security. Columbia University Press, New York, pp 46–86

Warner JF, Zeitoun M (2008) International relations theory and water do mix: a response to Furlong's troubled waters, hydro-hegemony and international relations. Polit Geogr 27:802–810

WB/ADB joint working paper on Future Directions for Water Resources Management in the Mekong river basin (June 2006)

Wolf AT, Yoffe SB, Giordano M (2003) International waters: identifying basins at risk. Water Policy 5(1):29–60

World Bank (2004) Water resources sector strategy: strategic directions for World Bank engagement. Report no 28114. http://www-wds.worldbank.org. Accessed 14 May 2008

Young O (1989) International cooperation: building regimes for natural resources and the environment. Cornell University Press, New York

Zeitoun M, Warner J (2006) Hydro-hegemony: a framework for analysis of trans-boundary water conflicts. Water Policy 8(5):435–460

Chapter 2
Politics or Development: Sharing of International Rivers in the South

Ashok Swain

Abstract In the 1990s, many agreements were drawn up in the South to share the international river basins. Noteworthy ones are Zambezi, Mekong, Jordan, Ganges and Nile rivers. These agreements are going through severe stress due to increasing water demand and climate change-induced uncertainties. In these cases, the riparian countries agreed for water sharing arrangements when they had hope for the further exploitation of the river resource. Possibility to acquire more water had led the political leaders to opt for the agreement as it provided political gain. However, these agreements are not worthy enough to initiate overall development in the basin as they lack the ability and motivation to make best possible use of the scarce water resources to meet future water challenges.

2.1 Introduction

Water has been called the oil of the twenty-first century. Global water consumption is rising steeply, and the lack of adequate supplies of fresh water is a problem in many parts of the world. Water is one of the most abundant elements of earth, covering nearly 70% of the planet's surface. However, only 0.003% of this huge volume is actually usable. Moreover, water availability is highly erratic in different regions of the world. More than 80% of the total global runoff is concentrated in the northern temperate zone, which houses a relatively small population. The volume of the rivers, which is the major source of the fresh water, is also unequally distributed among the countries of the less water available developing regions (Falkenmark 1993). Amazon River alone accounts for 80% of South America's average runoff.

A. Swain (✉)
Department of Peace and Conflict Research, Uppsala University,
Box 514, SE-751 20 Uppsala, Sweden
e-mail: Ashok.Swain@pcr.uu.se

J. Öjendal et al. (eds.), *Politics and Development in a Transboundary Watershed:*
The Case of the Lower Mekong Basin, DOI 10.1007/978-94-007-0476-3_2,
© Springer Science+Business Media B.V. 2012

Similarly, 30% of the total runoff in Africa originates from the Congo Basin. It is not easy to transfer water over long distances. In altering natural watercourses, many diversion canals have been and are being constructed to transport water from one part of the basin to another, to cities and to farmlands. But, due to geopolitical and economic reasons, it becomes really problematic to transfer water from one country to another or even one basin to another.

Thus, many long-distance water transfer proposals have remained in the planning stage for a long time. The Moscow City plans to revive an old scheme of water diversion from Siberian rivers to Central Asia, which was scrapped in 1986 by Gorbachev. Turkey has been planning for a long time with a large scheme to divert water from its GAP project on the Euphrates–Tigris River system to Gulf countries. In July 2001, American President George W. Bush publicly expressed his plan of persuading Canadian prime minister about piping Canadian water to the parched American Southwest, which was swiftly rejected by the Canadian environmental minister. High infrastructure and maintenance costs are the major deterrents for their execution. Such endeavours even become more problematic when the water has to be exported across several political units. This limits the possibilities for the political elites to find additional water resources to meet the ever-increasing water demand of their electorates.

Water tables are falling increasingly on every continent. Many countries in the South already face serious problems in meeting rapidly increasing water demands. Today, two or more countries share 263 major river basins. These shared basins cover more than 45% of Earth's land surface and support more than 40% of the world's population (Wolf et al. 2005). The increasing scarcity of water and the unequal and multilateral distribution of this resource pave the way for a greater number of international river water disputes. In the post–Cold War period, a number of commentators argued that the dependence of many developing countries on an external water supply might force them to re-orientate their national security concerns in order to protect or preserve such availability. The acute scarcity of water combined with the regional instability might lead to the use of force by the conflicting riparian states over the sharing of the river water resources. As early as the mid-1980s, US government intelligence services estimated that there were at least ten places in the world where war could break out over decreasing shared water (Starr 1991). Even recently, the UN Secretary General Ban Ki-moon in his address to delegates at the first Asia-Pacific Water Summit held in Japan in December 2007 warned, 'Water scarcity threatens economic and social gains and is a potent fuel for wars and conflict'. This conflict scenario brought the issue of water to the 'high politics'. Politicians as well as media came together to argue that the scarcity of water has replaced oil as the source of conflict. Many started seeing the greatest threats to the world's security coming from 'water wars'. However these 'water wars' are yet to be translated into reality.

In several cases, the competing riparian countries moved towards signing agreements rather than being engaged in armed conflicts. Shared water not only expected to increase competition and conflict, it can also contribute to build engagement and cooperation among the riparian states. Due to mutual dependence, the withdrawal or pollution of river water of one riparian state can potentially not only lead to the disputes but also bring cooperation in the basin. Particularly in the last two decades, several competing and disputing riparian countries have opted to reach formal agreements. In the twentieth century, 145 water-related treaties have been signed (Yoffe and Wolf 1999). Competing riparian countries of the Mekong, Jordan, Ganges, Nile and Zambezi rivers have signed agreements in the 1990s. The signing of the agreements on these important rivers in conflict-prone regions has been regularly used as examples to downplay the possibilities of 'water war' scenarios. Water is being increasingly painted as greater pathway to peace than conflict.

All of these agreements in Asia on major river systems in Africa and the Middle East in recent years have materialised in pursuit of acquiring more water resources in order to meet growing demand. Many of these agreements have been reached about how the river water should be shared to decrease the tension and create conducive political and economic climate to build new water projects on the river to increase supply. However, most of those agreements are perceived as unjust as upstream countries believe and unilaterally work towards controlling the flow of the rivers and using the maximum share. In some cases, where the downstream countries are more powerful like Egypt and Israel, they take all possible measures to challenge the upstream rights in order to secure their share. Brochmann and Gleditsch argue that treaties are important elements to spark more extensive cooperation among riparian countries (Brochman and Gleditsch 2006). But none of these critical river basins in Africa, Asia and Middle East have been able to establish a truly basin-based water management institution. In some cases, pressure and encouragement from the donor community have also facilitated the agreement process. However, these water agreements now face danger to their survival if they do not address the 'demand' side of the water issue as they fail to receive support from effective institutions for proper water management at the basin level. Due to political, economic and environmental limitations, the hope and aspiration of the agreements even fail to increase the 'supply' side. When countries bicker over the quantity of shared water and the threats of climate change become more apparent, the challenge for survival for sharing agreement is much higher (Wallensteen and Swain 1997). Allotted water in the existing sharing agreements in most of the cases is unable to meet the increasing demand. The scope of further augmentation of river water in the arid and semi-arid regions of the world is also getting limited due to possible impacts of global climate change.

2.2 Rivers of Conflict to Rivers of Cooperation?

2.2.1 Agreement over the Jordan River Between Israel and Jordan

The Jordan River rises on the slopes of Mount Hermon in Syria and Lebanon and moves to the south and passes through Lake Tiberias (Sea of Galilee) to empty into the Dead Sea. The Jordan River receives water from its major tributary, the Yarmuk River, whose catchment area lies in the Huran Plain and the Golan Heights as well as in some parts of Jordan. There are also other smaller tributaries to the Jordan River that originate in Jordan, Israel and the West Bank. From its origin to the entry of Lake Tiberias, the Jordan River is called 'upper Jordan', and the stretch between the Lake and the Dead Sea is called 'lower Jordan'. The River Jordan has a drainage basin of 18,300 km^2 situated in five political entities. The River Jordan carries a natural annual flow of 1. 470 km^3/year (Wolf 1995).

The struggle over the control of the Jordan River Basin is one of the most discussed subjects in the 'water conflicts' literature (Lowi 1995). It is the largest and longest river that flows in Israel. Moreover, it is the only river within Israel that has a permanent flow year round. Moreover, the other major rivers in Israel are contaminated with agricultural and industrial sewage, which makes the Jordan River the only natural and clean river in the country. In spite of its relative large size in Israel, Jordan River is actually a smaller river even in regional terms. With only 1,400 million cubic metres of usable annual flow, the Jordan River is the smallest major watershed in the region, compared with the Nile with 74,000 million cubic metres per year or the Euphrates at 32,000 million cubic metres per year. The Jordan River supplies Israel and Jordan with the vast majority of their water resource. As such, mostly Israel depends on water supply, which either comes from rivers that originate outside the border or from disputed lands. For the State of Jordan, the Jordan River supplies most of its needs. Jordan has a renewable annual water supply of 0.7 km^3/year, of which 50% is groundwater and 50% is surface water, mainly from the river Yarmuk. Lebanon and Syria are minor users of the water from the river Jordan.

The need for water and the continuing hostility between Israel and the surrounding Arab States has placed the Jordan River as a central bargaining chip since Israel's creation. The conflict over the Jordan River Basin surfaced immediately after the establishment of Israel. For Israel, as a young country, water seemed to be an integral part of its territory and a necessary resource for development. In the early 1960s, Israel used force to stop construction attempts by Arabs to divert water from the Jordan River to Syria. Control over the water bodies was one of the major reasons for the Arab–Israeli War in 1967, and the water issue also probably influenced Israel's decision to invade Lebanon in 1982. In 1964, Israel began to divert 320 million cubic metres of water per year through its National Water Carrier. Syria and Lebanon decided in 1965 to build canals to divert the Jordan's headwater, upstream of the National Water Carrier. Using armed forces, Israel destroyed the canal-building equipment and forced Lebanon to stop the construction of the project. In a series of attacks, Israel also stopped the Syrian project in July 1966. Though the Six-Day War started a year later, the contribution of water to the growing tension, which led to the war, cannot be dismissed. In the 1967 June War, Israel occupied the Golan Heights and brought under its domination all the headwaters of the Jordan River and a larger stretch of the Yarmuk River. The occupation of the West Bank also gave control of the lower Jordan Basin to Israel. The invasion of Lebanon and the creation of the 'security zone' in the south gave Israel greater control of the Jordan and Litani rivers (Elmusa 1996). Taking advantage of its new hydro-strategic position, Israel began to withdraw more water for its own use from the basin. By the early 1990s, Israeli utilisation of the basin's total discharge had reached more than 60% (Klein 1998).

On October 26, 1994, the prime ministers of Jordan and Israel signed a peace treaty, which brought an end to the state of war that existed for almost 50 years between the two countries. The peace treaty between Israel and Jordan included an Israeli commitment to provide additional water to Jordan. The treaty brought the normalisation of relations between the two countries, and its signing was closely linked with the efforts to bring peace between Israel and Palestine. Thanks to this 1994 peace treaty, in September 1995, an interim agreement between Israel and the Palestinians was concluded, where Israel for the first time recognised that the Palestinians have legitimate rights to West Bank water (Postel 1999).

With the 1994 treaty, Israel and Jordan agreed on allocations of water from the Jordan and Yarmukrivers and from Arava groundwaters. Israel has agreed to transfer to Jordan 50 million cubic metres of water annually from the northern part of the country. Both countries also committed themselves to build storage facilities to hold excess water from rain floods as well as build dams for river flow management. In terms of environmental conservation, Jordan and Israel agreed to protect the river from pollution, contamination or industrial disposal. Furthermore, the treaty brought forward a provision for establishment of a joint water committee to oversee issues regarding the quality of the water. However, this treaty was meant to be the beginning of a wider regional agreement, bringing Syria and Lebanon to the cooperation. But the disagreement on water sharing also became another hindrance in the peace negotiations between Israel and Syria in the late 1990s. Due to increasing water scarcity, Israel also itself started failing to honour water sharing with Jordan in 1998–2000 drought period. Though the conflict was contained, the bilateral water sharing treaty continues to be under tremendous stress due to decreasing water supply and difficult political situation in the basin.

Israel has also reached an agreement with Turkey in 2002, in which Turkey has promised to sell 11 billion gallons of water every year over a 20-year period. This quantity of water is estimated to be enough to satisfy about 7% of Israel's annual needs for potable water (3% of total water needs). The Water Purchasing Deal was signed on 6 August 2002 during a meeting between the Israeli Prime Minister Ariel Sharon and visiting Turkish Energy and Natural Resources Minister Zeki Cakan. The water, drawn from the Manavagat River in southern Turkey, will be transferred to the Israeli coast. However, the Israeli Ministry of National Infrastructure estimates that cost for super tanker shipment, unloading and conveyance facilities will be higher than what Israel presently spends on seawater desalination. This agreement has not been implemented yet due to high cost factor.

The situation in the Middle East has become quite complicated after the new Palestinian uprising in the occupied territories of Israel. The present hydro-political situation in the Jordan River Basin is of serious concern. Israel presently uses most of its available fresh water supplies. Occupation of the Golan Heights has provided Israel to control the upper catchment areas of the Jordan River and its tributaries. It has increased fresh water supply of this highly water-scarce country. The integration of this 'occupied' water to Israeli economy has brought further complications for basin-based cooperation over the sharing of the Jordan River water. For several years now, Israel has also given priority to build several desalination facilities.

In order to understand the core of the conflict between Israel and Jordan around the Jordan River, it is important to note the different perceptions of water between the two countries. Jordan, as part of the Arab world, perceived the water problem as part of the Arab–Israeli conflict. Therefore, for the Jordanians, water was always a matter of an Arab national pride. At the same time, Jordanian regime being prodded by its Western patrons agreed to sign the peace treaty as it expected Israeli support for resolving – or alleviating – its water shortages. By signing the peace treaty, Jordan hoped to receive Israeli support to build a water conveyance system bringing salt water from the Red Sea to the Dead Sea that would increase the water level of the Dead Sea and thus preserve tourism, agriculture and mineral extraction in the region. In 1997, the two states agreed to the Red Sea–Dead Sea Canal project, but the huge economic cost and also the opposition from the environmental groups have stopped the execution of the project.

2.2.2 The Ganges River Agreement Between India and Bangladesh

The Ganges–Brahmaputra–Meghna is one of the largest river basins of the world. This 1,634,900 km^2 of basin spreads over Bangladesh, Bhutan, India, Nepal and China (Tibet). The Brahmaputra River originates from near Lake Mansarovar and after flowing in China (Tibet) in easterly direction comes to north-eastern part of India and then to Bangladesh to merge with the river Ganges. The Meghna River originates in north-eastern region of India and then flows into Bangladesh to join the combine flow of the Ganges and Brahmaputra. The Ganges River originates on the southern slope of the Himalayan range, and on its way seven major tributaries augment its flow. Three of them – the Gandak, Karnali (Ghagara) and Kosi – run through Nepal, and they supply approximately 60% of the Ganges flow. After leaving Himalayas, the Ganges flows through India to enter into Bangladesh. However, the Ganges Basin itself is approximately one million square kilometres and is densely populated with 420 million people. The disagreement over sharing the dry-season flow of the Ganges came up between India and Pakistan in 1951, when India planned to build a barrage at Farakka, 18 km upstream from the East Pakistan (later Bangladesh) border. The proposed project included a 38-km link canal of 40,000 cusecs capacity to take off from the barrage to augment the waters of the Bhagirathi-Hooghly at the lower point to flush out the silt and to keep the Calcutta port navigable. In spite of Pakistani objection, India took unilateral decision to start the barrage construction in 1962.

After the independence of Bangladesh in 1971, various rounds of high-level official talks, formation of the Joint River Commission, visits of the heads of the governments followed but without bringing any long-term solution to water sharing disagreement.

The Farakka Barrage became operational in 1975 only for a 40-day trial period. Both India and Bangladesh signed short-term agreements for Ganges water sharing in 1977, 1982 and 1988. After 1988, they failed to reach an agreement due to decreasing availability of water at Farakka because of upstream withdrawals in northern India.

After years of unsuccessful attempt to reach any understanding, Bangladesh again brought up the issue in several international forums, which brought the bilateral relations to a further low. In the dry seasons, the average minimum runoff at Farakka was estimated in 1975 at only 55,000 cusecs. From which, India wanted to divert 40,000 cusecs with the help of diversion canal at Farakka, while Bangladesh demanded all 55,000 cusecs for its own uses. The increasing upstream withdrawal for the irrigation purposes in the Indian side had further reduced the dry-season flow at Farakka. From 1994, Bangladesh complained of getting only 9,000 cusecs in the acutest dry-season periods, which led to the assumption that the water availability at Farakka had come down to at most 49,000 cusecs in the dry seasons. This new figure created a further hurdle for the negotiators to reach a water sharing agreement (Swain 2004).

The change of governments in both India and Bangladesh in summer of 1996 brought new possibility for getting a bilateral water sharing arrangement for the Ganges River. The election of Sheikh Hasina, as the prime minister of Bangladesh provided the reason for her country's desire to improve the bilateral relationship with India. Coinciding with the change of government in Bangladesh, India also experienced the change of administration in New Delhi. The new United Front Government in India was interested to live up to their earlier image of friendly policy with the neighbours. After 8 years of accusation and counter-accusations, in December 1996, the prime ministers of India and Bangladesh signed the Ganges River water sharing agreement. Instead of usual short-term agreements to share the dry-season flow at Farakka, both the countries opted for a 30-year arrangement. This agreement was hailed as a landmark agreement for Indo-Bangladesh bilateral relations; it also provides another case in favour of 'water peace' proponents. Though the agreement wishes for an integrated management of the watercourses, the weakness of the treaty is that it does not include a clear mechanism to achieve this.

This treaty refers to some other water-related issues like flood management, irrigation, river basin development and hydropower generation for the mutual benefit of the two countries. The treaty stipulates that below a certain flow rate, India and Bangladesh each will share half of the water. However, the 1996 agreement has been based on the flow average of 1949–1988, but the real flow at Farakka in the 1990s was much less than that. To get a reliable figure, the water experts should have taken the average of the flow of last 10 years to the agreement. Unfortunately, the very first year of the treaty witnessed a severely low dry-season runoff in the Ganges River. With the help of the political support, the 1996 agreement withstood the challenge. However, the dry-season runoff of the Ganges River has improved since then, due to good rainfall in the upper basin areas and possibly increasing melting of snow in the Himalayan glaciers as a result of climate change.

The most important contribution of the 1996 Ganges Water Sharing Treaty is that it provided an encouraging bilateral environment for discussing and deliberating on a number of other river water sharing issues. Khaleda Zia government came to power in the second half of 2001. Her party, BNP, projects the anti-India position. But she also demonstrated her eagerness to move positively in her country's relations with India and not to spoil the stage, which has been set by the 1996 Ganges Water Sharing Treaty. After Sheikh Hasina's return to power in early 2009, bilateral relation between Bangladesh and India has improved further. But if the upper basin areas face a dry spell cycle, it will be a big challenge for the present agreement to satisfy both the parties. Bangladesh is particularly edgy as the reduced flow in the Ganges system has potentially wide-ranging socio-economic and environmental implications for Bangladesh (Mirza 1997). The long-term effect of faster glacier melting is going to be critical for the dry-season flow of the river and that will pose serious problems for the bilateral agreement to reallocate the decreasing water in the face of increasing demand (Swain 2010). Moreover, primarily due to India's strong reluctance, Nepal remains outside the water cooperation, reducing the possibility for a basin-based management of the Ganges River in the future.

Bangladesh signed the 1996 treaty with a hope to build a barrage on the Ganges at Pangsha, downstream of Farakka in Bangladesh. India supports this proposal and offers technical assistance as this will help to increase the water storage facilities of Bangladesh and will reduce its dependence on dry-season flow. However, Bangladesh also would like to increase the flow from the upstream by building storage dams along the Ganges tributaries in Nepal. Bangladesh argues on bringing Nepal into the arrangement but without success due to India's reluctance. On the other hand, India intends to divert the Brahmaputra River to Farakka Barrage via a canal through the territory of Bangladesh in order to augment supply. This 'Indian' plan fails to take off due to strong opposition from Bangladesh.

2.2.3 *The Mekong River Commission*

The Mekong River is the largest international river of mainland Southeast Asia. Six riparian states, China, Burma, Thailand, Laos, Cambodia and Vietnam, share the resources of this Mekong River. The river begins its flow from the Himalayas, then stretching down through Yunnan, south-western China province. Mekong, for a short length, forms the border between Laos and Myanmar before entering Laos and Thailand. Here, the Mekong itself creates a 900-km-long river border separating Laos and Thailand. The river continues its flow into Cambodia. Before draining into South China Sea, Mekong splits into nine-tailed dragon creating the Mekong delta in Southern Vietnam (Öjendal 2000).

After years of negotiations and failed attempts, in April 1995, at Chiang Rai, Thailand, four lower Mekong riparian countries, Thailand, Laos, Cambodia and Vietnam, came together and signed an agreement on Cooperation for Sustainable Development of the Mekong River Basin, giving birth to the Mekong River

Commission. The Mekong River Commission (MRC) was created with the support of the World Bank, the Asian Development Bank, various donor agencies and international organisations. The 1995 Agreement was based on two main principles: firstly, to 'reflect and protect the sovereign interest of each co-riparian', and secondly, 'to ensure the integrity of the final Agreement'. The Agreement calls for the creation of three permanent bodies, the Council (policy and decision-making), Joint Committee (coordination and technical expertise) and the Secretariat (executing branch). The Council possesses the authority to look at issues such as disputes on political grounds. It is composed of one senior member (at least of ministerial or cabinet level) of each of the participating riparian countries, with the chairmanship rotating among the riparian states for 1-year periods, and it meets annually; the Council is supported by a Joint Committee, which has regular meetings by senior representatives of a more technical nature. The executive body is the Mekong Secretariat, which executes the decisions of the Joint Committee, after having been approved by the Council.

Apart from the Council, the Joint Committee and the MRC Secretariat, the MRC structure also includes the National Mekong Committees. The National Mekong Committees are differently organised in the four member countries, each reporting to its own minister. The last structure attached to the MRC process is the Donor Consultative Group, which is composed of donor countries and cooperating institutions, and acts to provide a link between the donor countries and the MRC process and addresses concerns from both parties.

The Agreement stipulates that the member riparian states to cooperate in all fields of sustainable development, utilisation, management and conservation of the water and related resources of the Mekong River Basin in a manner to optimise the multiple use and mutual benefits of all riparian states. In terms of membership of the commission, the Agreement leaves the door open for any other riparian country to join, i.e., China and Burma, provided that they accept all the articles of the Agreement and are accepted by the existing members. When the 1995 Agreement was signed, it immediately came under criticism for its failure to include the unanimous principle of the Mekong Committee, e.g., other countries can veto diversion, dam or any project on both Mekong mainstream and tributaries that are considered as causing detrimental effects on them. The exclusion of this principle in fact gives individual riparian nations free hands to go ahead with their own individual plans.

The MRC primarily focuses on issues that affect the basin, such as hydropower, fisheries, agriculture and flood management where they are involved. In these sectors, it supports information exchange, cooperation and dialogue with the relevant agencies and institutions involved in planning projects, coordination between the actors involved, information gathering and studies on the environmental effects. The exchange of information continues to be conducted on a somewhat ad hoc basis. Added to the problematic of gathering and sharing information between the MRC countries is the issue of information exchanges with upstream countries, in particular with China. Under an agreement, China has now agreed to share information regarding the water level during the flooding season, between June 15 and October 15, from two stations in Yunnan Province in China (McKinney and Roeun 2002).

Overall, the 1995 Agreement is considered one of the most encompassing and holistic international water management agreements in the world, as it takes into account sustainable development, full utilisation of the rivers resources and the protection of the environment. What it lacks is comprehensive and definitive basin-based structure. It has been unable to deter the continued and unhindered construction of new dams in upstream China. An institution has been created but it does not include all the riparian states in the membership. Moreover, the cooperation structure is restrained by the political decisions taken by the member countries.

The 1995 Agreement outlines a wide scope of cooperation among four lower basin riparian countries that includes 'irrigation, hydropower, navigation, flood control, fisheries, timber floating, recreation and tourism' and other areas beyond those spheres. However, this cooperation spirit in the Agreement is only on pen and paper; what member states actually do is another matter. In the Mekong Basin, there are six riparian countries sharing the water resource. However, for its four lower riparian states, the river and its resources are crucial to their economies and their ways of life. The whole basin area is undoubtedly a contentious area. The six riparian countries share a difficult history with one another having frequently involved in armed conflicts. There are also border disputes between the riparian countries, and cultural differences. Moreover, there is also a clear disparity in the relative strengths among them. The upstream country, China, is the most powerful in the basin and also the least dependent on the resources of the river. Issues relating to the principle of sovereignty in the basin are very important, and actions that impinge on this principle are treated with scepticism.

There is no doubt that the lower basin countries came together in 1995, being urged by international donors. The hope was to receive financial and technical support to carry out large water projects in the basin. Unfortunately, the member countries of the Commission continue to be dependent on donor funds for their operations, as well as on the technical expertise provided by donors. Overall, the management of the basin area is not conducive to long-term sustainability. Due to climate change, increasing population and growing economy, four lower riparian countries might eventually abandon the collaborative process represented by the MRC and take unilateral actions.

2.2.4   The Zambezi River Commission

The Zambezi Basin covers eight countries: Zambia, Angola, Zimbabwe, Malawi, Tanzania, Botswana, Mozambique and Namibia. The River is 2,575 km long, and it rises in the north-west part of Zambia and flows to Indian Ocean at Beira in Mozambique. Zambia is the largest contributor to Zambezi Basin area, but Malawi is completely dependent on surface water resources. Besides Malawi, a large part of the populations in Zambia and Zimbabwe are dependent upon the Zambezi water. Angola, Namibia and Botswana have small proportion of their populations within the basin, but because of the future water demand, they still have strong interest in the basin management.

Amongst the most notable sectoral activities connected to the Zambezi River Basin are hydroelectric power generation, agricultural production, urban water supply, fisheries and tourism. An ongoing conflict between Zimbabwe and Zambia in the basin concerns over the expansion and usages of hydroelectric power production. Zimbabwean withdrawal of water from Zambezi River for its coal-fired Huangwe thermal station is also another issue of contention. There is also tension over the Zambezi River resources due to Zimbabwe's long-standing plan to pipe water from Zambezi (the Matabeleland Zambezi Water Project) to its drought-affected second city, Bulawayo. The water transfer to urban centres has not only brought tension with Zambia, it has also created opposing actors inside Zimbabwe. Sharing of the water has brought tensions between urban and rural people, large farm owners and marginal ones. The threat to Mozambique's water supply is not only limited to Zambia or Zimbabwe's water diversion from Zambezi. South Africa has a large water diversion plan, the Zambezi Aqueduct, to meet its water scarcity situation. South Africa intends to divert water over 1,200 km from the Zambezi River at Kazungula through Botswana to Pretoria.

In 1950s, Northern Rhodesia (now Zambia), Southern Rhodesia (now Zimbabwe) and Nyasaland (now Malawi) cooperated to construct the Kariba and Kafue Dams in the Zambezi River Basin. This, however, did not lead to the formation of a basin-based organisation to manage the river water resource. In 1987, the Zambezi River Authority (ZRA) was formed by Zambia and Zimbabwe concerning the utilisation of the Zambezi. The ZRA's mandate only covers that section of the Zambezi River forming the common border between Zambia and Zimbabwe, devoting largely to the operation and maintenance of the Kariba Complex. Although both these countries have differences over the operation of the Kariba Dam, they have been planning for some time to jointly construct another dam at Batoka Gorge upstream of Lake Kariba.

The Southern African Development Community (SADC) has taken a lead for the last one and half decades to facilitate better cooperation among its member states over shared water resources. In 1995, the SADC (all the Zambezi Basin states are the members of this organisation) signed a protocol establishing basic principles for the sharing of the region's water resources. For the Zambezi Basin, with UNEP support, the Zambezi Action Plan (ZACPLAN) was drawn in the 1990s. It aims to ensure sustainable utilisation of Zambezi water resources within a sound and balanced environment. Thanks to ZACPLAN, regional legislation and proposals for the establishment of a river basin commission have been developed. In spite of all these encouraging signs in 1990s, the Zambezi River Basin has not yet experienced the establishment of a River Basin Authority with the participation of all the riparian countries. After years of negotiation and pressure from the aid community, on 13 July 2004, all the riparian countries except Zambia have signed Zambezi Watercourse Commission Agreement.

So far, only Namibia, Mozambique, Angola and Botswana have ratified the Zambezi Watercourse Commission Agreement. Others, besides Zambia, have signed it but yet to ratify. Zambia has been refusing to sign as it argues that 75% of the Zambezi River Basin is in its territory which contributes 42% of total runoff. Zambia wants those aspects to be captured and factored in when it comes to water abstraction from the Zambezi River. Besides Zambia, the other major riparian

Zimbabwe, as Turton argues, is not showing much interest in basin-based planning because it may affect its predominant status within the existing Zambezi River Authority (ZRA) (Turton 1998).

In spite of hosting several international river basins, Southern Africa is very much of a water-scarce region. Major basin countries, those who have shown their willingness to be part of the proposed Commission, have very little interest in joint river management. Their consent to be part of this basin-based initiative is primarily guided by the expected international support to their planned unilateral water projects. Zambia and Zimbabwe are interested in building Batoka Gorge Dam about 50 km downstream of Victoria Falls, and that would include a 181 m high dam and would provide up to 800 MW of hydro capacity each for Zambia and Zimbabwe. The other projects of interests for these two major riparian countries are Devils Gorge and Mupata Gorge dam projects. The basin-based water cooperation in Zambezi Basin is still a distant dream in spite of agreements arrived in 1995 and 2004 due to Zambian opposition and Zimbabwean reluctance, as they do not expect to gain from the basin-wide cooperation.

2.2.5 The Nile Basin Initiative

Ten countries in the north-eastern part of Africa, Rwanda, Burundi, Congo, Tanzania, Kenya, Uganda, Eritrea, Ethiopia, Sudan and Egypt share the Nile River. From its major source Lake Victoria, the White Nile flows northward through Uganda and into Sudan where it meets the Blue Nile at Khartoum. Originating in Lake Tana in the Ethiopian highlands, the Blue Nile collects its flow from tributaries in Eritrea, Ethiopia and Sudan. From the confluence of the White and Blue Nile at Khartoum, the Nile flows northward into Egypt and on into the Mediterranean. Though it serves more than 150 million people, the Nile's average annual discharge of 84 billion cubic metres is modest in comparison to the other major river systems in Africa (Swain 2008).

Historically, Egypt is almost completely dependent on the waters of the Nile. The river provides more than 95% of the total water used in Egypt each year. Egypt has historically been the most powerful riparian country in economic and military terms, and taking advantage of this, it has been able to maintain its supremacy over the Nile water, and that has remained almost unchallenged until recently. Until the mid-1990s, several upstream riparian states were also plagued by political instability, internal conflicts and faulty national development strategies. Improved political and economic stability, growing populations and national demands for economic development have influenced the upstream Nile riparian countries now to develop their water resources to meet national development needs.

Water is very unevenly distributed in the basin. The White Nile upstream riparian states are well endowed with water resources, while Blue Nile riparian states suffer from scarce water supply. The White Nile upstream riparian states are currently determined to undertake various hydropower projects in the basin. They have also come together with Ethiopia in their opposition to 1959 Agreement between Sudan and Egypt and asking for basin-based sharing arrangement. These developments have posed significant challenges to the basin's current power relations, but the hydropower projects do not pose any threat to reduced water supply to Egypt, as they do not divert the water from system for agricultural or other purposes.

After signing 1959 Agreement, Egypt has always given priority to motivate Sudanese policy in favour of maintaining status quo on the Nile water sharing. Sudan is serious in constructing a number of large water projects in the Nile. With the help of Chinese and Arab funding, it has already built a large Merowe Dam, which is presently used for hydropower purposes, but in the future, it may include irrigation projects as well. The other ongoing project is the heightening of the old Roseires Dam to increase its hydropower production. However, Sudan is not happy in just generating hydropower; it has concrete plans to extend its capacity of irrigated agriculture, and that will take its water abstraction from the Nile above its stipulated quota in 1959 Agreement.

However, the real threat to Egyptian water supply comes from the Blue Nile Basin. The Blue Nile Basin contributes 86% to the total Nile flows reaching at Lake Nasser. Ethiopia, now with its improved economic and political strength and also better international standings, is starting to implement unilateral projects, which pose serious challenge to the long-established hydro hegemony of Egypt. Ethiopia has developed national water master plans for all the Ethiopian river basins with the help of international consultants. Several water projects have been already initiated unilaterally. In the Nile basins, besides several micro-dams in the highlands, Ethiopia has constructed a large hydropower dam, the Tekezze Dam, in the Atbara River. These projects have become possible due to favourable construction contracts offered by China. Ethiopia is hopeful of receiving further Chinese support for its planned irrigation projects in the Nile basins, including the controversial project of Tana–Beles. Emergence of China as a powerful alterative lender facilitates the possible unilateral actions of Ethiopia.

In the 1990s, the Nile River was considered, by many others, as a case of having high potential to induce inter-state conflict in its basin. However, thanks to the World Bank's initiative, a basin-wide cooperation, the Nile Basin Initiative (NBI), was launched in February 1999, of which all but Eritrea (participates as an observer) are members. In September 1999, NBI Secretariat was officially opened in Entebbe, Uganda. The NBI represents a transitional arrangement until the member countries agree on a permanent legal and institutional framework for sustainable development of the Nile Basin.

The Nile Basin Initiative is comprised of Council of Ministers of Water Affairs of the Nile Basin (Nile-COM), a Technical Advisory Committee (Nile-TAC) and the Secretariat (Nile-SEC). This NBI has developed a shared vision 'to achieve sustainable socio-economic development through the equitable utilisation of, and benefit from, the common Nile Basin water resources'. Joint development of Nile waters requires significant financial resources. The World Bank is coordinating an International Consortium for Cooperation on the Nile (ICCON), which aims to promote transparent financing for cooperative water resources development and management in the basin.

In spite of so much of hope and hype, in the last 13 years, NBI has not been able to transform the mindset of basin countries in thinking the Nile water development from a state-centric perspective to a basin-based strategy (Allan and Nicol 1998). The Nile Basin Initiative brought together all the riparian countries of the Nile River, where they officially expressed their desire to work for a joint initiative over the equitable utilisation of Nile River water resources. But, after years of meetings and deliberations, only in June 2007 the Nile Council of Ministers expressed their desire for establishment of a permanent river basin commission. Though basin countries formally agree for basin-wide cooperation, they continue to advocate and promote large-scale hydro projects unilaterally within their own territories. Furthermore, the basin countries have not taken any measure to reduce their dependence on the Nile River water rather their demand for water is consistently increasing. In spite of international community's support for cooperative water management of the Nile water, it has not been able to take a foothold; rather almost all the basin countries, particularly Ethiopia, Sudan and Egypt, have undertaken unilateral actions to protect their water interest. Emergence of China as a major player in African development process has provided alternative possibilities for Ethiopia and Sudan to raise financial and technical supports for their own water development projects.

Recently, the basin-based Nile Basin Initiative has come to a breaking point. After years of unsuccessful negotiations, in May 2010, seven upper riparian countries have come together in favour of Cooperative Framework Agreement. Ethiopia, Uganda, Tanzania, Rwanda and Kenya have already signed this Agreement, which is rejected by two powerful downstream, riparian countries, Egypt and Sudan. The main reason for their opposition is that the new framework asks them to share the Nile water with upstream riparian countries. The Cooperative Framework Agreement might soon get signed and ratified by other two upper riparian countries D. R. Congo and Burundi, and that will pave the way for the establishment of the Nile River Commission. However, the real challenge is to include Egypt and Sudan in the basin-based structure. However, their national interest prohibits them to take part, and without their participation, the Commission will not be in anyway capable to address the water sharing issues of the Nile River. Moreover, the unilateral actions of the upstream countries, particularly of Ethiopia and Sudan in the face of growing irrigation demand and climate change-related water supply uncertainties, have dashed any hope of achieving basin-based management in the near future.

2.3   Yet to Move Beyond Selfish National Interests

In the early 1990s, the major river basins like, Jordan, Ganges, Mekong, Zambezi and Nile were regularly branded as troubled spots and of having real potential of inducing armed conflict among their riparian states. The signing of agreements in all of these basins before end of that decade virtually altered the hypothesis altogether and brought the 'water peace' protagonists to the forefront of the debate.

Agreements arrived over the Jordan River and Ganges River in the mid-1990s were of bilateral nature, but they raised hope for culminating in basin-based cooperation involving other riparian states in the basins. In the Mekong Basin, the agreement among the disputing lower riparian countries to establish a River Basin Commission was certainly a positive development towards better management of this important river. It was expected that China and Burma would become part of this institutional framework sooner than later. The two important basins in Africa, Zambezi in Southern Africa and Nile in north-eastern Africa, even arrived at understandings to create basin-based water management structures.

All these agreements have been almost one and half decades old. Besides all the hypes, very little progress has taken place in establishing effective basin-based water management institutions in these basins. There has not been any process of importance at place in the Jordan and Ganges basins to bring in other riparian countries and move from bilateral sharing arrangement to joint basin-based development. Regional and national political considerations helped the signing of these bilateral agreements. India and Bangladesh in the new treaty managed to increase their existing share of the river flow. That helped the ruling elites to score political brownies vis-à-vis their oppositions. Moreover, both the countries opted in favour of the treaty, as the new improved bilateral environment would help them to build further water projects on the river. In the Jordan Basin, while regional politics was the main force behind this treaty, Jordan was also very hopeful of getting increased water share and implementing Dead Sea Canal project. Israel was making an investment in this agreement for larger and greater benefits, both politically and waterwise.

In the Mekong Basin, China not only is adamantly refusing to join Mekong River Commission but also has been pursuing large-scale unilateral dam buildings in the upstream. China has the political, economic and hydrological superiority in the basin to ignore other concerns and carry out its own project unilaterally. The dam building is not just confined to China; even member countries of the Commission are engaged in constructing large water projects unilaterally. In the Zambezi Basin, Zambia is similarly adamant in blocking any progress in establishing Zambezi Watercourse Commission. On the other hand, it has plans for a number of large dam projects together with Zimbabwe. The Nile Basin, where the international community, particularly the World Bank, has been claiming the credit since 1999 of creating a platform for basin-based water management, the ongoing standoff between Egypt and Sudan with other riparian countries shows shallowness in the claim. Egypt and Sudan do not want to forfeit their traditional importance over the Nile water for the sake of basin-based cooperation. It is a fact that none of these river agreements of the 1990s have led to the creation of a basin-based structure for joint cooperative water management.

Thanks to population growth and growing economy, all these five river basins are experiencing increased demand for fresh water. In the face of growing demand scenario, the supply side of fresh water, particularly runoffs in the river systems, also suffers from high fluctuation and uncertainties. The increasing threat of global climate change further complicates the future scenario. To meet the challenges from demand and supply sides, these five international rivers need basin-based water management institutions at place in order to avoid 'water conflicts' in the future. It is true that the agreements of 1990s have been able to hide water-based incompatibilities in these highly volatile regions for over a decade. However, their present form and progress are unsuitable to meet increasing water demand and climate change-induced supply uncertainties. It is necessary that countries in these basins need to come forward to manage their shared water resources in a cooperative and collaborative manner and actively work towards establishing effective river basin organisations. Successful basin-based organisations facilitate better integration of demand and supply, promote meaningful participatory processes and provide incentives for regional interdependencies. The inclusion of all basin states in joint cooperative activities open up new opportunities for achieving win-win solutions. Unfortunately, there is very little sign of that progress in these river basins rather there have been movements in the opposite directions.

The existing basin or sub-basin-based initiatives in Mekong, Nile and Zambezi basins came up in the 1990s due to pressure from international donor agencies, and they are still surviving with external aid and assistance. These initiatives unfortunately receive still very little contributions and support from the basin states themselves. It exposes the lack of sincere interest of basin states towards joint management of the shared river resources. Thus, it is important that researcher and policymakers need to take a close look at the ability and progress of these existing river agreements to face increasing water demand and the climate change related challenges in the near future.

References

Allan T and Nicol A (1998) Water resources, prevention of violent conflict and the coherence of EU policies in the Horn of Africa. Discussion paper, School of Oriental and African Studies, University of London, London

Brochman M and Gleditsch NP (2006) Conflict, cooperation, and good governance in international river basins. Paper presented at a meeting in CSCW working group 3, Environmental factors in civil war, PRIO, Oslo, 21 Sept 2006

Elmusa SS (1996) Negotiating water: Israel and the Palestinians. Institute for Palestine Studies, Washington, DC

Falkenmark M (1993) Regional water scarcity–a widely neglected challenge. People Planet 2(2):10–11

Klein M (1998) Water balance of the upper Jordan river basin. Water Int 23:244–48

Lowi M (1995) Rivers of conflict, rivers of peace. J Int Aff 49:123–44

McKinney M and Roeun V (2002) China agrees to share information on Mekong water levels. The Cambodia Daily, 2 Apr 2002

Mirza MMQ (1997) Hydrological changes in the Ganges system in Bangladesh in the post-Farakka period. Hydrol Sci J 42(5):613–631

Öjendal J (2000) Sharing the good: models of managing water resources in the lower Mekong river basin. Department of Peace and Development Research, Gothenburg

Postel S (1999) Pillar of sand: can the irrigation miracle last? W.W. Norton, New York

Starr JR (1991) Water wars. Foreign Policy 82(Spring):17–36

Swain A (2004) Managing water conflict: Asia, Africa and the Middle East. Routledge, London

Swain A (2008) Mission not yet accomplished: managing water resources in the Nile river basin. J Int Aff 61(2):201–214

Swain A (2010) Environment and conflict in South Asia: water-sharing between Bangladesh and India. South Asian J 28:27–34, April-June

Turton AR (1998) Water and state Sovereignty: the hydro-political challenge for states in Arid Regions. Paper presented at "Water Africa 99", Cairo, 30 May–1 June 1998

Wallensteen P, Swain A (1997) Comprehensive assessment of the freshwater resources of the world, international fresh water resources: conflict or cooperation? Stockholm Environment Institute, Stockholm

Wolf AT (1995) Hydropolitics along the Jordan river. Scarce water and its impacts on the Arab-Israeli conflict. UNU Press, Tokyo

Wolf AT, Kramer A, Carius A, Dabelko GD (2005) State of the world 2005 global security brief no 5: water can be a pathway to peace, not war. Worldwatch Institute, Washington, DC

Yoffe S, Wolf A (1999) Water, conflict and cooperation: geographical perspectives. Camb Rev Int Aff 12(2):197–213

Chapter 3
Politics and Development of the Mekong River Basin: Transboundary Dilemmas and Participatory Ambitions*

Joakim Öjendal and Kurt Mørck Jensen

Abstract The chapter takes a two-pronged approach. Firstly it reviews the fundamentals of the Mekong Basin system, providing a broad overview of the natural system and its basic water regime and from there defines the key developmental and governance challenges. Secondly, it performs a historical odyssey in order to assess which previous attempts have been made to regulate the system, and what we have learnt from them. At its core we find three contemporary tools developed to accommodate a sharpening regional politics with urgent development imperatives, all emanating from the MRC. These are the Water Utilization Project (WUP), the IWRM Strategic Framework, and the Basin Development Plan (BDP). They are scrutinized before we conclude that the MRC-agreement, as well as these three tools, have delivered valuable input to basin governance. Simultaneously we are pointing out that they have not provided the final solution for how to deal with the accelerating urge for exploitation of the system's natural resources.

The Mekong River basin[1] has attracted considerable attention due to a long and successful history of institutionalized river basin cooperation (Jacobs 1992), while at the same time experiencing recent challenges in terms of environmental and social change (Molle et al. 2009; cf. Hirsch and Jensen 2006). The basin is neither

---

*Sections of this chapter occurred in Phillips 2006, but have been revised, added, and updated.

[1] The Mekong River is also known as the *Dza-chu* in Tibet, *Lancang Jiang* in China, *Mae Nam Khong* in Thailand, *Mae Khong* in Laos, *Mekongk* in Cambodia, and *Cuu Long* in Vietnam.

J. Öjendal(✉)
School of Global Studies, University of Gothenburg, Box 700, 405 30
Gothenburg, Sweden
e-mail: Joakim.ojendal@globalstudies.gu.se

K.M. Jensen
Danish Institute for International Studies (DIIS), Strandgade 56, Copenhagen, Denmark
e-mail: kumoje@gmail.com

J. Öjendal et al. (eds.), *Politics and Development in a Transboundary Watershed:*
The Case of the Lower Mekong Basin, DOI 10.1007/978-94-007-0476-3_3,
© Springer Science+Business Media B.V. 2012

characterized by either water shortages nor open conflicts, but rather tensions are inserted through large scale threats to the ecosystem-based services, rapid modernization, enforced social change, broad and endemic poverty, a long subdued international water-related rivalry, and accelerating water utilisation. The institutional cooperation is in this context both noteworthy, and crucial (cf. Earle et al. 2010). It is also, at times, seen as patchy and incomplete (cf. Dore and Lazarus 2009).

More importantly in this context, is that it has been tasked to both deal with accommodating regional politics and to pursue IWRM-based development. As we saw in the Introduction, that is a difficult task, and it remains an open question as to what extent it has managed to do that, while also a number of regional key parameters are in flux. This chapter will serve to provide the background for Mekong development and politics, and to analyse how the process of pursuing these two tasks has played out. Hence, below, we will *first* briefly review the political economy of Mekong waters, *secondly* identify contemporary development challenges, *thirdly* assess some of the most concerted institutional efforts at addressing them, and *finally* analyse to what extent the contemporary institutional set-up is properly designed for solving the inherent dilemmas of transboundary water governance.

3.1 The Basin's Economic Geography

The Mekong-Lancang Basin covers a major territory, six countries and ranges from the highest of mountains in the Himalaya to the Pacific Ocean. On its route it passes distinctly different terrains and societies, and is subject to a variety of different political project and livelihoods system. Below we will review the most significant of these 'geographies', and their implications for regional politics and basin development.

3.1.1 The Physical Geography

The total area of the Mekong River basin is 795,000 km^2, and the river has an annual flow of approximately 475,000 MCM/year, depending on where it is measured. Largely located in the tropical zone of Asia, the basin is subject to monsoon rains, which fall highly asymmetrically over the year. This results in massive variations in seasonal flow of the river, where the wet season flow may amount to as much as 25–30 times that in the dry season (58,000 m^3/s, as compared to 2,000 m^3/s). The total length of the river is 2,600 km and it is the eight largest river in the world in terms of flow, the 12th longest, and the 21st in terms of river basin area. The source of the Mekong River lies in the Tibetan plateau in the south-west of China, and it runs through the six co-riparians to terminate in the South China Sea (Fig. 3.1).

Fig. 3.1 The lower Mekong River basin

The annual rainfall within the basin varies according to both geography and season. In the driest part of the basin (the Korat Plateau in north-eastern Thailand), the annual rainfall can be as low as 1,000 mm, whereas in more humid areas, it can reach three times this amount. By any standard the area is water-rich. Despite this,

Table 3.1 Physical statistics for the Mekong River basin (MRC 2003)

	China	Myanmar	Lao PDR	Thailand	Cambodia	Vietnam	Total basin
Area (km^2)	165,000	24,000	202,000	184,000	155,000	65,000	795,000
Catchment as % of country	38	4	97	36	86	20	N/a
Catchment as % of MRB	21	3	25	23	20	8	100
Average flow from area (m^3/s)	2,410	300	5,270	2,560	2,860	1,660	15,060
Rainfall as % of total in basin	16	2	35	18	18	11	100

in Cambodia and Laos less than 40% of the population has access to water of acceptable quality for potable use. In Thailand and Vietnam this ratio is distinctly higher, but in the Mekong Delta, less than 50% of the population has access to drinking water of an appropriate standard for drinking (MRC 2003:57).

The co-riparians to the Mekong River basin are China,[2] Myanmar (formerly Burma), Thailand, Laos, Cambodia, and Vietnam. The area of the basin is unevenly distributed over their territories (Table 3.1). In China, the river travels through Yunnan Province, covering a large part of that province, but a minor part of China as a whole. Myanmar has only a small area within the basin, while Thailand and Laos share the river as a border for a considerable stretch. In addition, a substantial part of the river passes through the inland areas of Laos, and the watershed covers most of that country. Cambodia is almost all within the watershed, whilst the basin covers a minor part of Vietnam (the Delta area), but this is exceptionally crucial in terms of economic value.[3]

The Mekong River basin is in many ways a pristine area with limited water consumption. No significant inter-basin transfers exist at present, and only two main stem dams are constructed; those and several more under construction are in the upper basin (see Chap. 9 Magee this volume). Although populated by some 75 million inhabitants, there are no major population centres (Phnom Penh far downstream being the only major city on the river), and no industrial centres. Even modern/intensive agriculture is limited to some parts of the Korat Plateau and areas of the Delta, in Thailand and Vietnam respectively. Hence, water quality is of a relatively high standard throughout the river basin, in general terms. A large number of tributaries exist to the main stem of the River, the most important including the Nam Num in Thailand, the Se San and the Ton Le Sap in Cambodia (see Chap. 6 Keskinen this volume), which have each been subject to controversial disputes in relation to various projects. Major plans exist in many areas of the basin for exploiting the resource in the future.

---

[2] The People's Republic of China, termed 'China' here for convenience.

[3] The central highlands of Vietnam and parts of the north of the country are also located within the Mekong River basin.

3.1.2 The Ecology and Livelihood Systems

Governance of the Mekong River system concerns access to water and its related resources, rather than water availability *per se*. One critical aspect involves the ecological system which sustains some 80% of the basin inhabitants. As Fox and Sneddon (2005:2) note:

> Critically, for millions of people, who live in the lowlands [of Cambodia], it is not the water alone that is the natural resource of greatest concern. Rather, it is the variability and complexity of an intact ecosystem – driven by annual flood pulse – that is the resource of immediate, and arguably highest, value.

There is relatively little agro-industry in the area (again, certain areas of Thailand and the Delta partly excluded); agriculture is at best semi-intensive, and there are a limited number of major plantations. Most of the population fish occasionally, and many engage in business related to fisheries, but the landless (the poor) rely on fisheries and other types of foraging for their very survival.

A special feature of the Mekong is that a large part of the plain is flooded, transferring nutritious silt on to the fields, inundating forests, and providing perfect spawning ground for many fish species. There are an estimated 1,200–1,700 species of fish in the Mekong, including a number of species threatened by extinction (such as the Irrawaddy dolphin and the Giant Catfish[4]; MRC 2003:57). The fish catch may be as high as two million tonnes a year and sustains 40 million people full or part time (MRC 2003:101; Chap. 4 Cooper, this volume). For instance, the Ton Le Sap provides one of the most efficient inland fisheries in the world and sustains 75% of the population in a large part of Cambodia with 75% of their protein intake. Biodiversity is immense, and two of the three last discovered mammals in the world have been found in this area. In 'normal' years, flooding is not a problem but rather a blessing for most people. However, the last decade has experienced several 'out-of-the-ordinary' flooding, with substantial material costs.

3.1.3 The Economic Geography of the Mekong Basin

The Mekong basin provides a multitude of opportunities for various stakeholders. Below follows a brief overview of the key rationales for resource utilization in various parts of the basin

The river travels through Yunnan, China, largely through a barren landscape with deep gorges and high altitude drops and only minor agricultural areas nearby.

---

[4] What possibly was the largest fish ever caught in fresh water was brought up in the Mekong system, in Thailand. It was a giant catfish measuring 300 kg, 'being the size of a Grizzly bear'. These giant catfish are rare these days and the discussion on the diminishing stock is directly connected to dambuilding and blasting of rock formations (Mydans, *The New York Times*, August 26, 2005).

Hence, there is limited exploitation of the water, with limited outtakes. Since the early 1990s, though, major mainstream dams have been built on the otherwise unexploited river. To date, two dams are completed, and several more planned (see Chap. 9 Magee this volume). Although they may have major environmental impact, no water is taken out of the basin (except through increased evaporation). For Laos, the Mekong and its water resources represent a hope for future prosperity, caught in the phrase describing Laos as the 'Kuwait of Southeast Asia', referring to its abundance of hydropower potential in relation to its population. The Mekong basin covers basically the entire country and the Laos territory is the area contributing with most water to the system (see Table 3.1). Currently, a few major dams are built, notably Nam Ngum, the first dam of international importance and concluded in 1965, while one more is newly constructed (Nam Theun 2), and several mainstream ones are debated, (notably Xayaburi, see below).

The Mekong River never crosses into Thailand, but the basin covers a considerable share of the country. Importantly, that part of the country is the most densely populated, agriculturally dependent and the poorest. This makes it dependent on the natural resources that the Mekong system offers (cf. Chap. 5 Mirumachi this volume). In this area, there are some irrigated land, and some intensive agriculture. The river crosses through central Cambodia and the basin covers a major part of the country. The river hits the plain in the northern part of the country, and the flooded area starts. At Phnom Penh, the mainstream is joined by the Ton Le Sap river. At the start of the rainy season, the level of the mainstream rises faster than in the tributaries, whereby the tributaries force the water to flow 'upstreams'. Most significantly, this takes place in the Ton Le Sap river, gathering water in the Ton Le Sap lake. This lake is up to five times as large in the wet season as in the dry season. Forests are inundated, plains are flooded, and the entire ecological system is defined by this annual 'pulse' (cf. Fox and Sneddon 2005). The river flows, finally, into Vietnam, where also the delta starts. This area constitutes a very complex and efficient fishery/agricultural area, which inhabits some 18 million people and produces more than half of the annual rice production in Vietnam. This low-lying area is constantly threatened by flooding as well as by saltwater intrusion.

3.1.4 The Political Geography of the Mekong River Basin Management [5]

As in all international river basins, the up/downstream relation is politicized and controversial, imbued with explicit or implicit power relations in terms of water utilization. The overall pattern here is that the upstream country – which always has an advantage over the downstream country(ies) – is China, which in terms of political

---

[5] Several of these controversial issues will be discussed at length below.

might is far superior to any of the downstream countries (or even the downstream countries together); the regional hegemon is thus also the upstream country, which results in extreme asymmetry in terms of power relations. Until recently, however, this has not meant much, since development activities pertaining to water in Yunnan province has been miniscule, leaving the downstream countries – i.e. Thailand, Laos, Cambodia and Vietnam – to negotiate the utilization of the river's resources between themselves.

This has been complex enough in itself. From a cold war induced artificial consensus, 'real' diverging interests resurfaced in the early 1990s. Thailand needs cheap energy (hydropower), more water to the modernized agriculture, and more water in the Chao Praya Basin stretching through central Thailand. Laos primarily wants to realize the hydropower potential, whereas Cambodia is best served by the conservation of the current hydro regime, including the seasonal flooding and maintained fish catch levels. Vietnam needs to build hydropower in the central highlands, as well as to protect the efficient agri- and aquaculture production in the Delta. These varying demands were successfully negotiated and codified in a framework agreement (the 'MRC Agreement' of 1995) in the early/mid-1990s and signed by the four countries in the lower basin.

In this constellation, Thailand and Vietnam are the major powers, but Thailand, with its economically more developed status as well as its upstream position, was a key player. Laos and Cambodia had minor influence, although it can be said that the overall mutual needs were at large respected in good faith (cf. Radosevich 1996). The drawback, it could be argued, is that the 'MRC Agreement' included a number of unresolved issues that are still subject to negotiations. The key of these may be the Water Utilization Project (WUP) – a process that has been financed by the World Bank, now in its seventh year of negotiations and with limited progress.

In addition to the above, China has, since a decade back, embarked upon a major dam-building program in the upper reaches of the Mekong. Already, significant impact is evident such as changing flow pattern and sediment trapping (MRC 2003:214), and it is likely that further dambuilding will exacerbate these indicative ecological problems (see Chap. 9 Magee this volume). Moreover, the water allocation formulas agreed upon in the 'MRC Agreement' by the four Lower Basin countries, is based on the then flows (i.e. before the dam-building in the upper reaches); if these are significantly altered, the agreement may turn obsolete, tearing down a promising cooperation. This would, in turn, take the process of sophisticated river basin management two decades backwards.

3.1.5 Development Challenges

The controversies on the Mekong are not water allocation per se. However, flows in quantitative terms nevertheless emerge as a key indicator on the condition of the river, the sustainability of fisheries, the risk for salinity intrusion etc. Below, we will outline the key challenges in terms of Mekong development (see Table 3.2).

Table 3.2 Key development challenges in the Mekong River basin

Key development challenges	Visible expression	Examples	Potential problems	Potential benefits	Resolving the dichotomies
Development intervention vs. Environ-mental integrity	For instance, building of major dams, blasting for improved navigation, building of bridges and roads	Nam Theun 2 (Laos); Yali Falls (Vietnam); Manwan, Daeshan (China); Yayaburi (Lao)	Minimal flow too low; maximum flow not sufficient; biological flows interrupted; reduced resource base	Improved infra-structure supporting economic moderni-zation and growth	(i) Ecologically sophisticated approach to 'development' (MRC Environment programme); (ii) Enhanced dialogue with environment movement; (iii) Only intervening if conditions fulfilled.
(In-)Equi-table utilization between up/down-stream riparians	Water consumption/ regulation (out-of-basin transfers; irrigation schemes, evaporation through damming)	Dam-building upstreams; Kon-Chi Mun; irrigation schemes (Korat plateau, vs Delta needs)	International tensions from unsolicited resource grabbing; sub-optimal development solutions due to insecurity	If successful, enhanced regional cooperation, accelerated economic growth, deepened and spirit of community	(i) Recognition of mutual needs throughout basin; (ii) Successful WUP process; (iii) Regional benefit sharing
Economic development vs. Poverty reduction	Basin development strategies determined by non-democratic elites, with thin public participation	Lack of public input to the MRC work; no local development gains from interventions such as Yali Falls, and Pak Moon	Rural under-development remains the same or even intensified; interventions supporting urban middle class only	Economic modernizati-on; economic benefit; regional growth	(i) Increased trans-parency and public participation; (ii) Successful BDP implementation; (iii) Enhanced pro-poor policies (e.g. Water management integrated in PSRP)

The most controversial aspect of Mekong development may be the possible dichotomy between *exploiting its natural resources* while at the same time *maintaining its ecological balance*. Originally, the great plan for the basin developed in the 1960s contained massive interventions with few social and ecological considerations (Öjendal 2000). Since then, both ecological awareness and political mobilisation has risen and from the early 1990s, codified in the 1995 Agreement, a more sophisticated thinking on these issues has emerged. A number of poorly planned/managed dam-building ventures (eg. Pak Moon, Pa Mong, and Yali Falls), and the protests and international tensions they triggered, served to put the issue of dam-building on the Mekong agenda. Both global (like International Rivers Network), regional (like TERRA), and local (like NGO-Forum) environmental NGOs have been highly efficient in turning this debate into a global concern. Pak Moon was for instance incorporated as a case study in the World Commission on Dams study and got a poor rating (WCD 2000).

As noted above, recently, the Lao government, with the support of the World Bank and others have built a major dam in Laos – Nam Theun 2 – which had been discussed for three decades. Export of hydropower provides a major item for Laos, and is expected to increase considerably, turning Laos into the 'battery of Southeast Asia'. Revenues from the Nam Theun 2 alone are projected to reach two billion USD for the next 25 years (Launey 2011). It is supposedly a project of the 'new generation' of hydropower dams where externalities are included in the estimated costs, redistribution of benefits to local stakeholders explicitly acknowledged, and environmental studies extensive. This has been followed by a push for several other large dams. However, critics abound and there is no doubt that it will make a major dent in the sensitive ecosystems and interrupt existing livelihoods. In a sense it is a major experiment as to whether big dams still can be built in complex ecological and cultural systems without severe damage being done and human rights violated in a major way. Or put differently, the ambition of the enlightened pro-interventionists is that with sufficient planning, the 'dichotomy' between infrastructure intervention and ecosystem integrity will vanish. Certainly that is how the Lao officials present it (BBC News, April 20, 2011).

For almost two decades, the building of mainstream dams in the Lower Mekong Basin seemed to belong to the trash bin of history. However, in spite of the Vietnamese call for dam-building moratorium, the Nam Theun, the building of several upstream mainstream dams in China, the increased hunger for energy in the region, and the global discourse on demand for 'clean energy' has turned the tables. Now, Laos has plans for up to nine mainstream dams, and Cambodia two, several more in the upper reaches in China, and dozens of major tributary dams are considered, littered over the lower basin (the lion's share in Laos, see Fig. 3.2) (International Rivers n.d).

The key to mainstream dams or not in the lower Mekong basin imay be the 1.260 MW/3.5 billion USD, Xayaburi dam. Rapidly rising to the fore, Laos has been vigorously pursuing this hydropower project, including closing a deal on financing arrangements with a Thai firm. The project would be financed by Thai banks and generate power mostly for sale in Thailand. The project has become an iconic battleground for the future of the Lower Mekong River. It has pitted activists, NGOs, villagers and the Thai and Vietnamese media against Thai commercial

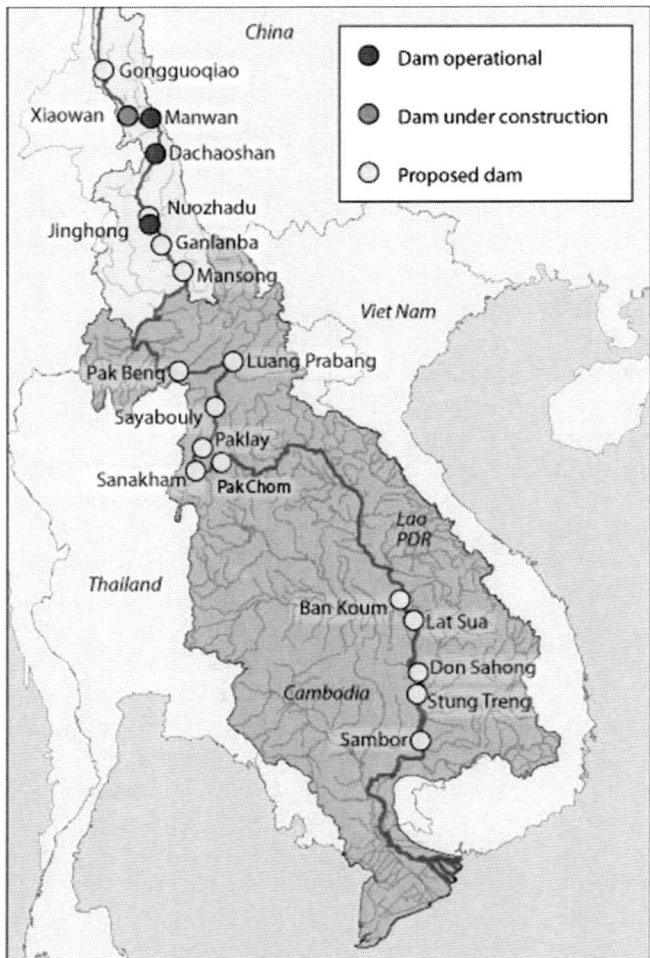

Fig. 3.2 Proposed and operational hydropower dams on the mainstream Mekong

interests and the Laotian government. In March 2011, 263 NGOs from 51 countries sent letters to the government of Laos and Thailand asking for the project to be shelved. Perhaps more significant has been the criticism of the project by Vietnam's official media and many Vietnamese scientists and environmentalists. Critics argue that the project could open the door for the ten other dams being considered for the Lower Mekong. The Xayaburi dam has also presented an opportunity (or made it impossible to avoid getting involved) for the MRC to engage in a deeply controversial development scenario with its broad knowledge based assessments and its political decision-making. The MRC member countries Thailand, Cambodia and Vietnam have expressed doubts on the soundness of the dam given its expected impact on fisheries and overall ecology particularly related to sedimentation.

During public consultations in early 2011 in Thailand, Cambodia and Vietnam there was no support for the dam (Hirsch 2011).

At a specially called meeting of the MRC in April 2011, no acceptance was issued and the matter was referred for decision-making at a ministerial meeting among the four countries to be held at a later date. The most recent chapter in the Xayaburi story was written at the last ASEAN Summit in Jakarta on 8 May 2011. In a closed side meeting between the Prime Ministers of Laos and Vietnam, the Laotian Prime Minister announced that Laos would temporarily suspend the project. It was agreed to engage "prestigious international scientists to seek firm scientific ground for future decisions" (Radio Voice of Vietnam 8 May 2011). This may be the final blow to the Xayaburi dam as the available knowledge and impact assessments of the dam is already considered to be scientific state-of-the-art knowledge. What remains is disagreement over the results of the existing scientific assessments. These differing interests appear to be embedded more in the political economy of water management than in the scientific information available.

A second key challenge in Mekong Basin management is to acknowledge the need to reach consensus on what 'equitable utilization' between up/downstream riparians might mean. Controversial issues here are minimum and maximum flow, water consumptions as in irrigation and evaporation from major dams, and plans on out-of-basin transfer of water. The latter is a highly contested issue. It is well known that some plans exist on transferring water from Lancang-Mekong into Yiangtse Kiang in China, and from Mekong to the Chao Praya river basin in Thailand. Both these are technically feasible and tempting, but highly controversial, and there is no evidence that they will be pursued within the near future.

Finally, the area is characterized by poverty. Although the countries are experiencing high economic growth, poverty in the rural areas is deep and structural, especially in Cambodia and Laos, but also in certain areas in Vietnam, and even Thailand. The poorest people are the ones most dependent on a functioning ecosystem, with rich fisheries, high biodiversity and all around access to lakes, streams, wetlands and forests. Modernization tends to produce high yielding

The Xayaburi hydropower project would be the first such project on the Mekong mainstream downstream of China and would be capable of generating electricity, mainly for export to Thailand. The Xayaburi dam is located approximately 150 km downstream of Luang Prabang in northern Lao PDR. The dam has an installed capacity of 1,260 MW with a dam 810 m long and 32 m high, and has a reservoir area of 49 km^2 and live storage of 225 Mm^3. The primary objective of the Xayaburi dam project is to generate foreign exchange earnings for financing socio-economic development in Lao PDR. The developer is Ch. Karnchang Public Co. Ltd. of Thailand who negotiated a tariff agreement with EGAT in July 2010. MRC Media Release 19 April 2011.

interventions, like dams, returning investment to the state or to the urban middle class (or even national or transnational companies), whereas it is the rural poor that typically have lost security and income (cf. Fox and Sneddon 2005; Cowan and Shaw 2003). It is rare that major projects manage to balance loss and risk from the side of the poor with benefits experienced by the better-off (cf. McCully 1996).

So, above we have a view of the development challenges at stake, and we can state that the challenges can all to be framed in the IWRM-transboundary field of tension as understood in the Introduction to this volume. Below we will review tools at hand for addressing these governance challenges.

3.2   Attempts at Water Governance in the Mekong Region

3.2.1   General Overview

The cooperation around water resources in the Mekong basin is often mentioned as one of the most successful in the third world (Jacobs 1992; Radosevich 1996; cf. Phillips et al. 2006). It has roots back into the 1940s, and it has survived such violent events as the Vietnam War and the Khmer Rouge in Cambodia. It has also persisted in spite of structural impediments such as the cold war induced division among the countries and historical animosities between the riparians (cf. Öjendal 2000). This question has been handled by various institutions; the original Mekong Committee (MC) dating back to 1957, changed into the Interim Mekong Committee (IMC) as a result of the Khmer Rouge and the subsequent civil war, triggering a Cambodian absence, and then, most importantly into the "1995-Agreement", creating the Mekong River Commission (MRC). The latter is a modern agreement, based on international water law and with certain attention on sustainable development and participatory mechanisms (Radosevich 1996; Öjendal 2000; Makim 2002; Chap. 8 Hirsch this volume). This is also the point of departure for contemporary attempts at balancing international interests with development imperatives.

3.2.2   The MRC Agreement 1995 – Contemporary River Basin Cooperation

[The MRC as] the latest chapter in the effort to harness the mighty Mekong River attests to the proposition that the 'Mekong spirit of cooperation' will continue to be a model among multinational efforts in international river basin development (Radosevich 1996:263).

Following the end of the cold war, the solution to the Cambodia conflict, and the subsequent recognition of the government of Cambodia in 1993, many expected a rapid return to the original statutes of the Mekong Committee. However, rather late in the re-enactment of the institutionalised lower basin cooperation, Thailand announced that it was not pleased with the previous agreement and wanted to renegotiate. In retrospect, this is hardly surprising: in the 1960s, there was a united front against communism, thus forcing consensus. Now, Thailand was an upstream country which hardly could be expected to undersign an agreement where all interventions needed full regional consensus. Instead, intensive negotiations ensued (Radosevich 1996; Makim 2002).

After 3 years of negotiations, a new deal for the Mekong management was struck and the MRC Agreement was signed in April 1995. There were substantial changes as compared to the 1957 and 1975 agreements/principles:

- The entire Agreement is based on, and emphasizes repeatedly, the idea of 'sustainable development',[6] and 'Environmental and Ecological Balance' (MRC Agreement 1995, Article 3).
- In line with modern views on water management, the Agreement covers not only water allocation, but also 'irrigation, hydropower, navigation, flood control, fisheries, timber floating, recreation and tourism, in order to optimize the multiple use and mutual benefits for all riparians...' (MRC Agreement 1995, Article 1);
- The previous *de facto* veto right is abolished, reducing upstream commitments in hard terms. The right to veto has been replaced by three levels of restrictions pertaining to various circumstances. Various interventions require: notification, prior consultation, and agreement by the joint committee (MRC Agreement 1995).
- The only distinct level of restriction – agreement by the joint committee – refers to inter-basin transfers in the 'dry season'. This can be done without the agreement, however, if there is a 'surplus'. These two concepts – 'dry season' and 'surplus' – were never defined in the agreement and have led to prolonged negotiations (MRC Agreement 1995, Article 5).
- Natural minimum and maximum flows are protected, so as to prevent saltwater intrusion and to preserve the natural water regime around the Ton Le Sap (MRC Agreement 1995, Article 6).

Flows and water allocation are never mentioned in quantitative terms in the Agreement, but rather all difficulties are collected in Article 26, which nicely sums up the unresolved issues at the time:

> The Joint Committee shall prepare and propose for approval of the Council, inter alia, Rules for Water Utilisation and Inter-Basin Diversions pursuant to Articles 5 and 6, including but not limited to 1) establishing the time frame for the wet and dry seasons; 2) establishing the location of hydrological stations, and determining and maintaining the flow level requirements at each station; 3) setting out criteria for determining surplus quantities of water during the dry season on the mainstream; 4) improving upon the mechanism to monitor intra-basin use; and 5) setting up a mechanism to monitor diversions from the mainstream.

In all, the new agreement returns power to the individual nation-state, away from a regional regime, and repositions power upwards in the system. At the same time,

---

[6] The formal name of the agreement is: 'Agreement on the cooperation for the sustainable development of the Mekong River Basin.'

however, it increases the demands on the sustainable utilization of the river's resources, the institutional capacity of the Secretariat, and the 'soft' demands on policy harmonization. The consequence of this, and even more directly of Article 26, was that a number of programmes were initiated and put to work under the new Mekong River Commission's rather large secretariat.

3.2.3   Implementing the MRC Agreement

In the 1995 Agreement the four countries in the Lower Mekong Basin agreed to cooperate in the sustainable development of the water and related resources of the Mekong for *"An economically prosperous, socially just, and environmentally sound Mekong River Basin"*. The MRC was established as the focal point for the cooperation to assist the member countries in implementing the agreement through the provision of shared information, technical guidance, and consultation. The agreement inter alia called for the formulation of a Basin Development Plan to promote development and prevent inappropriate use of the basin's water and related resources. A number of development programmes have been launched to implement the MRC Agreement. The MRC Secretariat is responsible for these programmes arranged under an umbrella 'Mekong Programme'. There are three 'core' and a number of sector programmes. The three core programmes are: (i) the Water Utilization Program (WUP) now followed by the Mekong IWRM Project; (ii) the Basin Development Plan (BDP) which is supposed to "promote the coordinated development and management of water and related resources" (MRC Agreement 1995, Article 2); and (iii) the Environment Programme (EP) which is crucial for the 'sustainability' aspect of the MRC agreement.[7] As an overall framework for implementing the Agreement and its related development programmes, the MRC has prepared 5-year Strategic Plans that stipulates the strategic direction and scope of work of the organisation. The fourth of such Plans are now in operation covering 2011–2016.

3.2.4   The Water Utilization Project and the Mekong IWRM Project

The *WUP* was implemented from 2000 to 2008 with funding from the Global Environment Fund (GEF) through the World Bank (WB). The project objective was to support improved water management in the basin through agreed water utilisation

---

[7] The descriptive aspect of the review of the three core programmes below is collected from the MRC material.

procedures and guidelines. In order to accomplish this objective, the WUP produced a *decision support framework* consisting of a suite of basin wide models, a knowledge base including a comprehensive hydrological modelling package, and preliminary environmental impact assessment tools. A number of case studies of 'hot spots' in the basin were also carried out in order to: (i) demonstrate trans-boundary environmental management; and (ii) provide examples of alternative development scenarios illustrating the economic, environmental and social benefits as well as costs to each MRC country of certain developments (MRC 2004).

The *decision support framework* provided a hydrological overview of the basin and contributed widely to the understanding, function and potential for develop-ment in the Mekong Basin. It is now institutionalised under the MRC's Information and Knowledge Management Programme (IKMP) that manages all MRC data, information, models, impact assessment tools etc. in support of the many MRC programmes. The more analytical, strategic and operational application of basin development scenarios is now the responsibility of the BDP programmes as will be explained below.

The *WUP* also focused on establishing the basis for MRC member country agreement on the procedures and technical guidelines for water use as specified in the 1995 Agreement. The rules, procedures and guidelines were meant to: (i) provide a mutually agreed basis between Member Countries for defining the sustainable limits for water-related basin development at any point in time; (ii) establish a monitoring system to ensure that the sustainable limits are not being exceeded, and (iii) provide information that will help guide the future development of the "Mekong Basin" (MRC Agreement 1995 Article 6 and 26). The WB worked closely with the MRC on the *WUP* to develop the required regional knowledge on the dynamics of the Mekong (e.g. basin wide models and a framework for trans-boundary analysis of development scenarios) to support the establishment of a regional enabling framework though agreed procedures and technical guidelines for water use.

- Procedures for Data and Information Exchange (approved 2001)
- Procedures for Notification, Consultation and Agreement (approved 2003)
- Procedures for Water Use Monitoring (approved 2003)
- Procedures for Maintenance of Flows on the Mainstream (approved 2006)
- Procedures for Water Quality (approved 2011)

As could be expected, the 'soft' issues of information-sharing, procedures for notification and monitoring have been easier to negotiate than the 'harder' ones on water flow and water quality. The procedures for maintenance of flows on the mainstream and for water quality proved much more difficult and required lengthy discussions and negotiations between the four countries. The reasons for this may include insufficient awareness and the complexity of the procedures and their related technical guidelines. Also, strong perceptions on national sovereignty may have led to the suspicion that the procedures on water flow and water quality are restraining and regulatory mechanisms. However, by 2011 all five procedures and their technical guidelines for their implementation have been agreed by the four countries.

There is widespread recognition that the WUP has produced tangible outputs and managed to stimulate the Lower Mekong cooperation for almost a decade, not least because of the intense MRC member country dialogue and negotiations, including periods of disagreement and national reflection, on the adoption of the water use and water quality procedures. An independent evaluation in 2007 acknowledged that the *WUP* had been "… successful in meeting the legal requirements, commitments and expectations set out in key documents …" (Independent Evaluation 2007). The only exception was the water quality procedures adopted in 2011. The evaluation stressed the need to finalise procedures and guidelines and to extend these beyond the MRC Secretariat and National Mekong Committees and engage line agencies and key ministries in decision making on Mekong water and related resources.

3.2.5   IWRM as the Strategic Framework for Mekong Development

The 1995 Mekong Agreement does not mention IWRM. However, in the design and implementation of the Water Utilization Programme (and its successor the Mekong IWRM programme) and the Basin Development Programme the IWRM approach to water management is adopted as the overall policy framework that acknowledges the complexity of relations between water, environment, economic development and livelihoods. Similarly, IWRM is centre stage in the MRC's last two Strategic Plans for the periods 2006–2010 and 2011–2015 and also in the "IWRM-based Basin Development Strategy for the Lower Mekong Basin" agreed by the MRC member countries in January 2011.

In some countries and organizations such as the MRC, IWRM has gained ground in legislation, policies and water action plans. Although it is difficult to speak against IWRM at its highest aspirational level, some critics argue that the concept may be best at maintaining a glossy rhetoric narrative of how water ideally should be managed while in reality water is managed according to power and politics serving the interest of powerful economic stakeholders. In this vein it is argued that IWRM is too ideal for the complex political economy of water where goals and stakeholder interests may be "frequently if not always antagonistic" (Molle 2008).

Although at national level there are some differences in the way IWRM is interpreted and adopted, all MRC member countries subscribe to IWRM as the guiding approach to water management in the Mekong. The main feature of consensus is the agreement that any infrastructural or other development on or linked to the river should maintain the 'triple bottom line' of sustainability. Accordingly, water management efforts must analyze alternatives to address the potentially conflicting goals of economics (financial), environmental, and social issues. This approach has been applied to a large extent by the MRC in the operational work of the MRC's Basin Development Plan particularly in the assessment of various Mekong development scenarios (MRC 2010). Also, the recent independent Strategic Environmental

Assessment of mainstream dams on the Mekong sponsored by the MRC Secretariat can be seen as belonging to the overall IWRM approach (MRC 2009). In addition, the IWRM approach has introduced a significant component of participation and stakeholder consultation and involvement in many MRC activities, particularly within the framework of the Basin Development Plan and the Procedures for Notification, Prior Consultation and Agreement (PNPCA) process for the Xayaburi dam on the Mekong mainstream. With these achievements it would appear that the MRC with its technical Secretariat has managed to move beyond the IWRM rhetoric and policy and actually developed knowledge packages based on at least some of the intentions of IWRM. The extent to which the MRC countries are using IWRM-based knowledge in their decision making on critical developments such as mainstream dams will test the political acceptance of such knowledge and science. Similarly, the extent to which MRC countries share knowledge and engage with their respective constituencies and stakeholders is also a test to the stakeholder and governance imperative of IWRM.

3.2.6 MRC's Basin Development Plan

The 1995 Agreement mandates the MRC to engage in basin development planning in order to enable member countries "...to promote, support, cooperate and coordinate the development of the full potential of sustainable benefits to all riparian states". Thus a BDP programme was launched in 2001 to support and facilitate a joint planning process that aims at facilitating regional development through effective, sustainable and equitable management of the Mekong water and related resources. The BDP continues the grand 'planning tradition' within river basin management, although by other means.

The first phase of the BDP (2001–2006) established a basin-wide planning process and methodology rather than a blue-print grand master plan with a list of projects. The result was a *rolling basin planning cycle* for the Basin including a knowledge base and planning tools. In its substance it was largely driven by IWRM as the holistic and inclusive approach to management and development of the basin as a hydrological unit. It introduced a participatory dimension through a bottom-up approach to basin planning based on stakeholder involvement in ten sub-areas of the Basin.

The second phase of the BDP (2007–2010) produced tangible outputs that placed the BDP programme at the centre stage of the MRC cooperation. The relevance of basin development planning was underpinned by the actual developments on the river including extensive plans for dams and hydropower projects on the Mekong mainstream and tributaries. The BDP also expanded its participatory approach from consultations at the sub-area level to a series of stakeholder consultations involving regional and national NGOs and civil society.

Taking into account future plans and projects on the river, the BDP2 prepared assessments of a number of *basin-wide development scenarios*. The scenarios included an analysis of water use and demand and the balancing of trade-offs between different uses of water. In the assessment of the scenarios economic developments were being weighed against agreed environmental and social objectives and criteria. The scenario assessments provided a basis for discussions on levels of water resources development and their associated environmental and social impacts leading to what was defined as the *development opportunity space* (MRC 2010). The development opportunity space is a hydrology based concept. The 'space' for development is defined by parameters for water storage and extraction so that each country will be able to share the benefits while at the same time the flow of the river is maintained at a level that will prevent further saline intrusion in the Mekong delta.

The development opportunity space is at the core of the second and probably most important output of the second phase of the BDP namely the *IWRM-based Basin Development Strategy*. The Strategy is seen as a milestone in Mekong cooperation and is defined as a development and management 'statement' by the MRC member countries that addresses current and future challenges and opportunities. It presents a roadmap to implement the Strategy according to which regional and national planning will be linked in order to register and assess specific development investments (projects) in the basin. The Strategy presents the prior consultation for the Xayaburi mainstream dam in Laos as first 'test case' for the implementation of the Strategy.

The Strategy starts out with a baseline account of how the Basin's water resources are currently being developed and managed, and it identifies likely development trends. This baseline includes also the Upper Mekong i.e. China. The Strategy has a *development* and a *management* part. The water resources *development* part addresses the short and long term (up to 2060) development opportunities and risks within the hydrological 'space' of the River. Here the focus is on hydropower and irrigation as the main drivers of economic development. There are assessments of the many proposed tributary dams as well as the planned and more controversial 11 mainstream dams in Laos (9) and Cambodia (2) (Fig. 3.2). The assessments are guided by the 'triple bottom line' of the IWRM approach and include environmental, fisheries, bio-diversity and social impacts from hydropower and irrigation developments. Finally, the likely climate change impacts are also included. The *management* part of the Strategy seeks to establish management arrangements for water-related sectors such as fisheries, navigation, floods and droughts and, wetlands. It supports the integration of regional and national water management procedures through capacity building programmes. Finally, the Strategy emphasizes enhanced regional and national stakeholder participation "…respecting community and wider popular participation approaches in each country." (MRC 2011:17)

The BDP will continue into its third phase by 2012. The focus will be in the implementation of the agreed roadmap in the *IWRM-based Basin Development Strategy*. After 8 years of the MRC's work on basin planning there is a strong sense of realism of what can be addressed and what can be done through basin development planning. The realism is based on lessons learned over the years, many completed or planned hydropower projects throughout the Basin, and a strong sense of national sovereignty expressed through the development interests of each member country. The geo-politics of the region has its own dynamics that may not necessarily be easily aligned with the ideal IWRM approach to the Basin as a hydrological unit.

3.3   Geopolitics and the IWRM Approach to Basin Development – Conclusion

The WUP (and its successor Mekong IWRM Programme) and the BDP have been constructed to deal with two key issues of the MRC Agreement: the allocation of water resources and the profile of development interventions. The two programmes have also been the strongest drivers of the IWRM approach to basin management. In combination they would realize the MRC Agreement and go a far way in bridging the divide between national interests and geopolitics and the one hand the ideal IWRM notion of management and development of the Mekong basin on the other. Let us look a bit closer to what it has achieved and which overall consequences it has rendered.

3.3.1   The Role of IWRM in a National-Interest Driven Mainstream Dam Evolution

Development of the Mekong's water resources seems to be guided by two very different imperatives (which also is the inspiration of this entire volume). The dominant one is the national imperative founded on each member state's sovereignty and national development priorities. This imperative is ultimately framed by the political economy of each nation state, and the Mekong is largely seen as an open water regime (a common of sorts) that is there to be tapped more or less freely according to the economic needs of each country. The less dominant imperative is the river basin imperative that is concerned about managing and developing the transboundary waters of the Mekong as an interconnected hydrological entity with a diverse ecosystem with an imperative of pursuing sustainable development.

In the MRC context, IWRM has been adopted as *the* accepted management approach. One of the key developmental contributions from the MRC has been the assessment of various Mekong development scenarios that includes planned hydropower facilities/dams on tributaries and the mainstream. In this manner, the river basin and IWRM imperative has been merged with the national imperative i.e. the dams planned unilaterally by each MRC member country. In other words, a merge of the national interests versus the 'common good'. As long as the 'common good' remains an open regime and national interests are satisfied, there is little room for disagreement. But once the regime closes, antagonistic interests are likely to emerge.

The early plans for the Mekong were encapsulated in the notion of a Mekong cascade of dams and hydropower projects on the mainstream (cf. Öjendal 2000). Such a scheme, devised in the late 1960s, would have changed the Mekong from being a free flowing river to an engineered river consisting of a series of lakes. Because of war and conflict over the following decades, these plans were shelved. They have now reemerged as all the countries in the lower Mekong are experiencing spectacular economic growth and the need for energy to fuel further growth is growing.

The MRC has adjusted to these realities not only through its BDP assessment of development scenarios but also through its recently established Hydropower Programme. The Programme has made attempts to apply IWRM's 'triple bottom line' in an assessment of the likely impact from planned mainstream dams on the environment, fisheries and people's livelihoods as well as the economic and poverty alleviation priorities of the MRC Member Countries. By supplementing its assessment work with a good IWRM practice and dialogue with MRC member countries and private hydropower investors, the MRC Secretariat has tried to take on its transboundary mandate as a river basin organization supporting the 'common good'. This automatically brings the MRC into the elusive area of international 'soft' law, regional agreements and transboundary governance. In this terrain the MRC is unlikely to solve the dilemma of IWRM-based water governance and the 'common good' at the basin level versus national water governance without a corresponding political will on national and local level.

The Xayaburi dam has been pursued as a necessary development imperative for Laos to bring foreign exchange earnings and economic growth to alleviate poverty. It is a classic case of national interests taking priority over the regional common good. The events around Xayaburi dam have probably been the biggest test to MRC's ability to have a constructive role in basin governance. So far the MRC has passed the test. At the technical level, relevant IWRM-based knowledge was provided and managed through various impact assessments. At the governance level, the MRC agreed procedures were followed and although far from perfect (see Chap. 8 Hirsch and Chap. 4 Cooper this volume) there was an element of participation of the public through views from and various stakeholder groups and local consultations in Thailand, Cambodia and Vietnam. Although being so far a 'good practise' case, the Xayaburi story has also exposed the geopolitical challenges that a regional organization like the MRC will have to face in future. The Xayaburi is not a one-time shot in history. Even if mainstream dams are not built, the open water regime of the Mekong could be closing if all (or even some of) the planned tributary dams will be constructed and some of them may have more negative impacts than the Xayaburi. The role of the MRC in tributary dams is likely to be much less than in mainstream dams. Although tributary dams with 'significant transboundary impact' should go through the MRC prior notification procedures, it remains a grey area as reaching agreement on 'significant transboundary impact' is difficult and open for interpretation. And as most tributaries are not transboundary (with the exception of the 3S tributaries: Sesan, Srepok, Sekong), national sovereignty is not at stake.

Although the current MRC priority attention to hydropower appear justified as an underpinning of its intergovernmental character, it has given national interests priority over those of a wider array of stakeholders, the public in general and vulnerable groups in particular. The Xayaburi case illustrates how national interests are reduced to 'state interests' in national economic growth without any inclusion of alternative voices promoting a more sustainable and participatory approach.

In line with the participation and governance components of IWRM, the MRC has been trying to make up for its deficit in participation and dialogue with civil society, NGOs and interest groups. Since 2007 there have been considerable developments in stakeholder consultations. Also, the dialogue between the MRC and the public (mostly expressed through international, regional, Thai and Cambodian NGOs) flows more freely now. Although much progress is done, the MRC and its Secretariat is challenged by the political realities in the region where structural differences exist between countries with regard to the nature of participation and mechanisms for public voices to be heard. The question remains as to how the 'responsibility' for involvement of and participation and engagement with the public should be divided between the MRC as a regional and multilateral organization and each nation state. It would appear that in the case of national government decisions on large-scale infrastructure projects such as the Xayaburi, having both transboundary and national importance, at least part of the 'responsibility' for dialogue and consultation with the public and stakeholders in Laos rests with the Laotian government. This would be good IWRM practice.

3.3.2 Final Note

Clearly, the Mekong cooperation has entered a new phase, and we have seen three milestones conforming that. *Firstly*, the BDP and the WUP – long anticipated tools – have been concluded and is delivering regional public goods. While they may not be the ultimate solution as once believed, they are crucial cornerstones in the MRC cooperation matrix, and will be delivering (if nothing else) procedures and *modus operandi* for basin cooperation for a considerable period of time. *Secondly*, the Xayaburi story opens up a new field of negotiations and accommodation, which may be determined by regional high politics, rather basin management tools, and only to a very little degree by references to IWRM practices. *Thirdly*, the Mekong issue is rising in political status: special ministerial meetings are called (as a result of Xayaburi stalemate), summits are held (such as that at Hua Hin in June 2010), and China is entering the scene, being prepared to talk about river basin development.

The MRC-Agreement and all its procedures and knowledge infrastructure has made the political process of accommodation much easier. Conflicting national interests have been couched in agreed procedures, transparent processes, and knowledge management. This is conducive to accord and accommodation. In spite of its rhetoric, it is not obvious that it has managed to pursue IWRM-development as well. It is of course an irony of history that the first mainstream dam in the lower Mekong is pursued just as the IWRM-based tools reviewed above are set to work. Or, perhaps, that is just what could be expected.

References

Cowan E, Shaw BJ (2003) Mekong River development – whose dreams? Which visions? Water Int 28(2):268–275

Dore J, Lazarus K (2009) De-marginalizing the Mekong River commission. In: Molle F, Foran T, Käkönen M (eds) Contested waterscapes in the Mekong region: hydropower, livelihoods and governance. Earthscan, London, pp 357–382

Earle A, Jägerskog A, Öjendal J (eds) (2010) Transboundary water management principles and practice. Earthscan, London

Fox C, Sneddon C (2005) Flood pulses, international watercourse law, and common pool resources: a case study of the Mekong lowlands. WIDER research paper 2005/22

Hirsch P (2011) A new geopolitics of mekong dams? Bangkok Post, 1 May 2011. http://www. bangkokpost.com/news/investigation/234766/a-new-geopolitics-of-mekong-dams. Accessed 21 May 2011

Hirsch P, Jensen KM (2006) National interests and transboundary water governance in the Mekong. AMRC, University of Sydney, Sydney

Independent Evaluation (2007) Independent evaluation of the water utilization start-up project, final report. 28 Dec 2007. Mimeo, Vientiane

International Rivers. http://www.internationalrivers.org/southeast-asia/mekong-mainstream-dams. Accessed 17 June 2011

Jacobs J (1992) International river basin development and climatic change: the lower Mekong of Southeast Asia. PhD dissertation, Department of Geography, University of California

Launey GD (2011) The stalled battle for the future of the Mekong River. BBC news, 20 Apr 2011. http://www.bbc.co.uk/news/business-13139470. Accessed 21 May 2011

Makim A (2002) The changing face of Mekong resource politics in the post-Cold War era: re-negotiating arrangements for water resource management in the lower Mekong River basin (1991–1995). AMRC, working paper no. 6

McCully P (1996) Silenced rivers – the ecology and politics of large dams. Zed Books, London

Mekong Agreement (1995) Agreement on the cooperation for the sustainable development of the Mekong River basin. The Kingdom of Cambodia, Lao People's Democratic Republic, the Kingdom of Thailand and the Socialist Republic of Vietnam, Apr 1995

Molle F (2008) Nirvana concepts, narratives and policy models: insight from the water sector. Water Altern 1(1):23–40

Molle F, Floch P, Promphakping B, Blake DJH (2009) The 'Greening of Isaan': politics, ideology and irrigation development in the Northeast of Thailand. In: Molle F, Foran T, Käkönen M (eds) Contested waterscapes in the Mekong region: hydropower, livelihoods and governance. Earthscan, London/Sterling, pp 253–282

MRC (Mekong River Commission) (2003) State of the basin report. Mekong River Commission, Phnom Penh

MRC (Mekong River Commission) (2004) MRC integrated basin flow management, 2004–2008. Mekong River Commission, Vientiane

MRC (Mekong River Commission) (2009) Strategic Environmental Assessment (SEA) of Mekong mainstream dams. Study commissioned by the MRC, May 2009. Mekong River Commission, Vientiane

MRC (Mekong River Commission) (2010) Assessment of basin-wide development scenarios; MRC basin Development Plan Programme phase 2, Nov 2010

MRC (Mekong River Commission) (2011) IWRM-based basin development strategy for the lower Mekong basin. Mekong River Commission

MRC (Mekong River Commission) homepage (2005) www.mrcmekong.org/programmes/wup/ wup.htm. Accessed 1–13 Oct 2005

MRC Media Release (2011) Lower Mekong countries take prior consultation on Xayaburi project to ministerial level. MRC media release 19 Apr 2011. http://www.mrcmekong.org/MRC_news/ press11/Lower-mekong-coutries-take-prior-consultation19Apr11.html. Accessed 21 May 2011

Öjendal J (2000) Sharing the good – modes of managing water resources in the lower Mekong River basin. PhD dissertation, Department of Peace and Development Studies, Göteborg University

Phillips DJ, Daoudy M, Öjendal J, Turton A, McCaffrey S (2006) Trans-boundary water cooperation as a tool for conflict prevention and for broader benefit-sharing. Ministry for Foreign Affairs, Stockholm

Radio Voice of Vietnam (2011) Vietnam fosters cooperation with Laos and Thailand. http://english.vovnews.vn/Home/Vietnam-fosters-cooperation-with-Laos-and-Thailand/20115/126419.vov. Accessed 21 May 2011

Radosevich GE (1996) The Mekong – a new framework for development and management. In: Biswas AK, Hashimoto T (eds) Asian international waters – from Ganges-Brahmaputra to Mekong. Oxford University Press, Bombay

WCD (2000) Dams and development: a new framework for decision-making. World commission on dams. Earthscan, London

Chapter 4
The Potential of MRC to Pursue IWRM in the Mekong: Trade-offs and Public Participation

Rachel Cooper

Abstract The Mekong River Commission (MRC) is committed to implementing IWRM in a development context that has a strong infrastructure focus. This raises questions about how to balance the goals of IWRM (economic efficiency, social equity and environmental sustainability) and the interests and power capabilities of the actors involved in the development of water resources. This chapter focuses on two key areas: the identification and negotiation of development trade-offs and public participation. Development of water resources involves trade-offs. In the Lower Mekong, the key trade-off is the development of hydropower versus fisheries which could bring local livelihoods and economic development into conflict. Public participation is extremely important within IWRM. However, there are issues of power and access that need to be mediated in order to allow a range of stakeholders to play a role in the development debate in the Lower Mekong. Through exploring these two areas, this chapter discusses some of the challenges the MRC must navigate if it is to play a strong role.

The Mekong River Commission (MRC) is committed to implementing IWRM in a development context that has a strong infrastructure focus. This raises questions about how to balance the goals of IWRM (economic efficiency, social equity and environmental sustainability), and the interests and power capabilities of the actors involved in the development of water resources. This chapter focuses on the potential of the MRC to pursue IWRM in two key areas: the identification and negotiation of development trade-offs and public participation. Development of water resources involves trade-offs. In the Lower Mekong, the key trade-off is the tension between hydropower and fisheries, which potentially places economic development and local livelihoods in conflict. Public participation is extremely important within IWRM.

R. Cooper (✉)
Newcastle University, Newcastle Upon Tyne, Tyne and Wear NE1 7RU, UK
e-mail: rachel.cooper@newcastle.ac.uk

J. Öjendal et al. (eds.), *Politics and Development in a Transboundary Watershed:*
The Case of the Lower Mekong Basin, DOI 10.1007/978-94-007-0476-3_4,

However, there are issues of power and access which need to be mediated in order to allow a range of stakeholders to play a role in the development debate in the Lower Mekong.

Through exploring these two areas, this chapter discusses some of the challenges the MRC must navigate if it is to play a strong role. Challenges include informing debates, planning and decision making through the generation and dissemination of relevant knowledge; increasing engagement with relevant actors such as line agencies and civil society; facilitating open dialogue between actors with different power capabilities; and mediating issues of power and access. The chapter explores the lower Mekong's development context and the role of the MRC, before turning to the organisation's potential in the areas of trade-offs and public participation.

4.1 The Development Context

Poverty reduction and economic development are key goals of the states of the lower Mekong. Utilisation of the Mekong's water resources is perceived as one strategy to realise these goals. Hydropower is a key development priority and driver of national interests (Lang 2006; ADB and World Bank 2006). There are 261 planned or proposed hydropower projects in the Mekong region (King et al. 2007). In 2007, plans to dam the lower Mekong mainstream were controversially resurrected with proposals for 11 dams. The rationale for increased hydropower development centres on poverty in the basin, hydropower's contribution to national development and the growing energy demand in the region.

Government development plans and strategies link hydropower development and poverty reduction. For example, Lao PDR's *National Growth and Poverty Eradication Strategy* (NGPES) argues that development of the state's considerable hydropower potential will bring extensive benefits and is therefore integral to the national development framework (Government of Lao PDR 2003). The MRC recognises that Member States 'regard the development of their hydropower potential as an integral component of their policies to continue…economic growth and so gradually eliminate poverty that is still all too prevalent within the Lower Mekong Basin' (Bird 2008c).

The lower Mekong has an enormous hydropower potential: 30,000 MW, of which only 5% has been developed (World Bank 2004). Energy demand is predicted to rise dramatically in the region (King et al. 2007). In this context, hydropower is a saleable commodity, and both the governments of both Lao PDR and Cambodia have expressed the desire to become the batteries of Southeast Asia, exporting electricity to their neighbours. Thailand and Vietnam intend to meet their growing energy needs through importing electricity produced by dams in their Mekong neighbours (Klopper 2008; Middleton 2008). Reductions in 2009 in the amount of electricity the Government of Thailand plans to import due to the global financial crisis have resulted in delays to some tributary hydropower projects in Lao PDR (MRC 2009b). This has been interpreted by the MRC as a short pause before demand increases again (MRC 2009b).

The rationale for hydropower development largely presents the lower Mekong River as unutilised. This masks the on-ground reality in which the lower Mekong River plays a vital role in the lives and livelihoods of the basin's population of roughly 60 million. The majority of the basin's population are dependent on the Mekong's water resources to some extent for agricultural, domestic and fishing purposes. The lower Mekong is one of the world's most productive fisheries valued at approximately US $2.5 billion annually. Fisheries resources play an integral role in livelihoods. Between 64% and 93% of rural households are involved in fisheries to some extent (Coates et al. 2003). Moreover, fisheries provide various types of employment: direct and non-direct, full- and part-time and commercial and non-commercial. They are also extremely important for food security. Consumption of fish and aquatic resources provides between 47% and 80% of animal protein intake, and in 2006, consumption of fish and other aquatic animals was estimated at 2.6 million tonnes (Hortle 2007). The subsistence and nutrition roles of fisheries enable people to engage in other means of employment (Hortle 2007). Tensions exist between the role of fisheries in the basin and the proposed hydropower developments (see Sect. 1.3).

The four states of the lower Mekong have adopted IWRM nationally through their water laws and policies, and collectively through their cooperation in the MRC. However, IWRM is a multi-faceted approach. Drawing on the 3E principles of efficiency, equity and environmental sustainability, IWRM argues that 'waters should be used to provide economic well-being to the people, without compromising social equity and environmental sustainability' (Varis *et al.* 2008: 146). In contrast to earlier water management paradigms, IWRM demands the incorporation of social and environmental concerns. Consequently, if water resources are used for economic purposes, this should not be at the expense of the goals of social equity and environmental sustainability. In the context of the lower Mekong, it is unclear how to balance these three goals. Civil society actors have expressed concerns that in infrastructure developments to date, economic considerations have dominated over social and environmental ones (e.g. International Rivers 2008). It is important to note that IWRM is not anti-water infrastructure development per se, but is a process that argues that social and environmental considerations should be given equal weight.

4.2 The Role of the MRC

The MRC was established in 1995 with the signing of the *Agreement on the Cooperation for the Sustainable Development of the Mekong River* (hereafter the Mekong Agreement). The organisation is mandated to 'cooperate in all fields of sustainable development, utilization, management and conservation of the water and related resources of the Mekong River Basin' including amongst others hydropower and irrigation (Mekong Agreement 1995: Article 1); development of a basin development plan and a role in coordinating development (Mekong Agreement 1995: Article 2); and protecting the environment and ecological balance, including from harm resulting from development plans (Mekong Agreement 1995: Article 3). The MRC is comprised of three bodies: the Council, the Joint Committee (JC) and the Secretariat. The Council is the governing body of the MRC comprised of a ministerial representative from each member state. The JC acts as a management body implementing policy decisions and supervising the Secretariat. The MRC Secretariat is the operational arm of the organisation, implementing various programme activities and providing technical and administrative services. National Mekong Committees (NMCs) in each state are responsible for coordinating the MRC's work at the national level.

The MRC formally adopted IWRM in 2005. Congruent with the goals of IWRM are the MRC's vision (an 'economically prosperous, socially equitable, and environmentally sound Mekong River Basin') and overall strategic goal ('More effective use of the Mekong's water and related resources to alleviate poverty while protecting the environment' (MRC 2006: 24)). The MRC's *Strategic Directions for IWRM in the Lower Mekong Basin* identified eight strategic priorities including economic development and poverty alleviation, environmental protection and integration through basin planning (MRC 2005). The organisation is also committed to strengthening the IWRM capacity and knowledge base of MRC bodies, line agencies and other stakeholders (MRC 2006).

The MRC focuses on basin-wide projects and plans, transboundary projects and national projects with significant/cumulative basin-wide implications (MRC 2006). The Basin Development Plan (BDP) Programme is a key programme tasked with implementing Article 2 of the Mekong Agreement. The BDP's second phase (which commenced full operation in 2008) will develop a rolling IWRM Basin Development Plan articulating a common development vision for the basin and providing directions for a rolling planning framework which will bring the basin perspective into national planning and vice versa (Hang and Lennaerts 2008). The IWRM Basin Development Plan comprises three elements: an IWRM-Based Basin Strategy, Development Scenarios and a Project Portfolio.

In terms of implementing IWRM in the current development context, the MRC has a number of possible roles. This chapter will focus on the MRC's role in terms of facilitating trade-off identification and negotiation and public participation. The impacts of hydropower and other water use changes in the basin will impact on fisheries and livelihoods. The MRC as a knowledge-based organisation believes that it can inform debate and planning in the negotiations of trade-offs in the lower Mekong basin (Interview MRC Official 07/08[1]). In 2008, the MRC renewed its commitment to public participation. The MRC has identified facilitating dialogue with, and between, different stakeholders as one of its key contributions to the ongoing development debate in the region (MRC 2009c).

---

[1] Interviews were conducted in the lower Mekong region between February and July 2008 with a wide range of actors. They are referenced in the text by their category and the month/year they were conducted.

4.2.1 Relevancy, Mandate and Business-as-Usual

The MRC is seeking to play a strong role in IWRM in a context where its role, mandate and relevance have been questioned. Developments in the basin such as Chinese dam development of the upper Mekong, and the transboundary issues surrounding the Yali Falls dam in Vietnam, resulted in civil society actors questioning the relevance and role of the MRC. The MRC's absence or partial engagement cultivated a perception of an organisation disconnected from important decisions and debates (Interview Environmental NGO Representative 06/08). Donors became concerned that financial support to the MRC had not resulted in a successful or engaged river basin organisation (Backer 2007; Hirsch et al. 2006). Concern was also expressed that the Member States viewed the MRC as a hindrance, perceiving it as a regulatory institution imposing rules (ADB and World Bank 2006). These statements illustrate the range of concerns surrounding the MRC's role.

Debate about the MRC intensified in 2007 with the revival of plans to dam the lower Mekong mainstream. The MRC was initially silent in the development debate arguing that it was an intergovernmental organisation whose mandate was to serve Member States' needs. Civil society actors, such as the Thai NGO Towards Ecological Recovery and Regional Alliance (TERRA), accused the MRC of abdicating responsibility and not fulfilling its primary duty of protecting the river (TERRA 2007). Donors also expressed concern about the role of the MRC in hydropower and about the impacts hydropower could have on fisheries and livelihoods in the region (MRC Meeting 11/07). There is a lack of clarity about the role of the MRC in hydropower. A range of actors, including donors, want to see the MRC take a more active role in debates and decision making in the basin.

The MRC operates at the intersection of a wide range of actors including Member States, line agencies, donors, civil society actors, researchers and scientists. These actors have different positions concerning the nature of Mekong development and the role of the MRC. Donors perceive the 1995 Mekong Agreement as a framework primarily concerned with the environment and livelihoods and premised support for the organisation on it playing a role in supporting and facilitating informed decision making (Hirsch et al. 2006; Lee and Scurrah 2009). In contrast, Member States argue that the Mekong Agreement is a development agreement (Interview State Official 06/08). Civil society actors, such as the Thai People's Network for the Mekong and Rivers Coalition of Cambodia, argue that the MRC should call for a moratorium on dams until scientific evidence has been collected, publicly disseminated and consensus reached. A key challenge for the MRC is to address the various claims in this debate over its role, clarify its mandate and engage this wide range of stakeholders in its work.

The MRC's commitment to IWRM and sustainable development has also been questioned. Within wider global debates about IWRM, it has been argued that the adoption of IWRM can allow actors to continue business-as-usual under a new label, gain access to new funds or increased legitimacy and repackage old projects (Biswas 2004; Molle 2009). These concerns have been expressed in relation to the MRC.

Despite a commitment to sustainable development, commentators have argued that engineering and economic concerns still predominate, and that the aims of Mekong cooperation were always, and still remain, hydropower development (Lang 2006; Sneddon and Fox 2006). This raises the questions of national interests and government commitment to the MRC.

The MRC is an intergovernmental organisation: it cannot act over and above its Member States. Consequently, national interests play a large part. Officially, the NMCs are the representatives of the national interest. However, national interests more commonly reflect the interests of other ministries, such as those of Finance, Planning and Energy, which play a larger role than the NMCs in national decision making, and are focused on the economic benefits of developing the Mekong (Lee and Scurrah 2009). If the MRC is to play a strong role in the basin, it must navigate these national interests and bring in the transboundary, basin and social and environmental perspectives into negotiations and decision making.

The MRC instigated a more active role in the Mekong hydropower development debate in late 2008. A number of key activities were initiated under the Initiative on Sustainable Hydropower (ISH), including a Strategic Environmental Assessment (SEA) of lower Mekong mainstream dams, studies on fish migration and support for the Member States in implementing the Procedures for Notification, Prior Consultation and Agreement. BDP2 fast-tracked work on development scenarios, including a mainstream dams scenario, and strengthened its stakeholder engagement with a range of actors including research networks. The MRC increased activities related to extending public participation and engagement with a wider range of actors, including line agencies and civil society.

Increased activities and visibility in the basin's hydropower debate suggests that the MRC is attempting to pro-actively engage with this important issue. The MRC has recognised that the hydropower development debate represents a critical time for the organisation. If it cannot exert tangible influence, then questions about its relevance, impact and effectiveness will continue (MRC 2009a). This suggests that it may not have been a case of business-as-usual but that the organisation experienced a period of limbo and required time to adjust to a rapidly changing development context. The increased engagement of the MRC is partially focused on two important areas: trade-offs and public participation.

4.3   Trade-offs in the Mekong

As we saw in the Introductory chapter, IWRM is an inherently political process. It envisages a balance between the 3Es, but it is not clear how to achieve this. Conflict between the demands of the three goals is also likely. Various actors will favour different goals and argue that more weight should be given to one over the others. Consequently, IWRM involves the meditation of conflicts of interest (Jonch-Clausen 2004). As the three goals may be antagonistic, trade-offs are necessary. Trade-offs are difficult to achieve as they involve a number of actors and competing interest. They are also controversial as they involve sacrifice on the part of some actors. It is important to note that different actors have different power capabilities, and that political bargaining and decision making are necessary. In order to make informed choices, good quality information and knowledge is required that is disseminated amongst various stakeholders.

In terms of trade-offs, the MRC has largely identified its role as that of knowledge generation and provision which can inform debate and decision making. There are a number of challenges associated with this role including how to ensure that MRC knowledge does inform and impact debate and decision making, and reaching out to decision makers such as various line agencies that are traditionally removed from the MRC's work.

4.3.1 The Nature of Trade-offs in the Mekong

The discussion of trade-offs in the lower Mekong basin recognises that there are costs and benefits to development, and that mechanisms are needed to reconcile competing interests and values. This requires knowledge, capacity, engagement with decision makers and public participation. In the lower Mekong development context, the key trade-off, identified by a number of actors, is hydropower and fisheries: hydropower brings benefits, but also dis-benefits in terms of blocking fish migration routes (ADB and World Bank 2006; MRC 2009d). The MRC aims to identify and negotiate trade-offs within the framework and principles of IWRM.

The MRC refers to the balancing of the 3Es as IWRM's triple bottom line (MRC 2006). A triple bottom line approach evaluates any proposed development in terms of its contribution to economic efficiency, social equity and environmental sustainability. Consequently, economic, social and environmental outcomes are all 'seen as part of the development benefit/dis-benefit...not that there is a simple trade-off between economic benefit, on the one side, and socio-environmental costs on the other' (Hirsch 2006a: 24). In the current hydropower debate, the MRC is using the language of triple bottom line and benefits/dis-benefits. However, the debate is largely framed by a range of actors, including civil society organisations, in terms of the three goals being antagonistic, with economic benefits on one side and socio-environmental losses on the other, such that an increase in hydropower will lead to a decrease in fisheries and livelihoods.

Hydropower development brings economic benefits such as increased government revenues. These benefits are located at the national level. Government strategies, such as Lao PDR's NGPES (2003), envisage these economic benefits 'trickling-down' and increasing socio-economic development through, amongst others, increasing the amount of money the government can spend on poverty reduction programmes. However, hydropower development can have social and environmental dis-benefits. In the case of the lower Mekong, this is mainly in terms of hydropower's impact on fisheries and the livelihoods that depend on them.

The Mekong's fisheries are one of the most valuable, productive and diverse inland fisheries in the world. Around 120 species of fish are commercially traded in the region (Coates et al. 2003). As shown above, fisheries play a unique and important role in the basin in terms of livelihoods. The threat that large infrastructure such as dams pose to the Mekong's capture fisheries is widely recognised (Poulsen et al. 2004; ADB and World Bank 2006). The impacts of existing hydropower dams on the lower Mekong's tributaries, including a decline in the abundance of fish, have already been documented (MRC 2003). Proposed lower Mekong mainstream dams pose a particular threat to fisheries because they will block fish migration routes. A large number of Mekong fish species are migratory, migrating up- and downstream to breed and feed. More than 70% of the total fish catch in the Mekong (roughly 1.8 million tonnes) is dependent on long distance fish migration, with the Mekong mainstream acting as a migration corridor (Dugan 2008). Blocking fish migration routes impacts not only on the fish themselves, but on livelihoods. Reducing the amount of fish reduces the availability of food for people, reduces food security and has an economic impact on poor people and their livelihoods (Interview Mekong Fisheries Scientist 06/08). That hydropower dams generally, and the proposed lower mainstream dams specifically, will impact fish migration is widely accepted amongst Mekong actors including civil society and organisations such as the MRC.

4.3.2   The Debate over Fisheries Mitigation

The possibilities for mitigating fisheries impacts are contested. An independent Expert Fisheries Group convened by the MRC in September 2008 reported that existing mitigation measures, such as fish ladders and lifts utilised in other basins, are not suitable for the Mekong due to the large quantity and volume of fish species and migrations (Dugan 2008). In contrast, Mekong government representatives have argued that there is currently not enough knowledge about fisheries in the Mekong to be able to conclude that technological mitigation is not possible, and therefore, more studies are needed (MRC Meeting 09/08; MRC Meeting 05/08). It has also been suggested that it may be necessary to look at other mitigation measures and not just fish in isolation (MRC Meeting 09/08).

Other possible mitigation options include developing aquaculture and reservoir fisheries, providing alternate income streams for affected communities and changing the location of proposed hydropower dams, constructing them higher up in the basin or on the tributaries to reduce impacts (Dugan 2008; MRC Meeting 09/08). However, measures such as aquaculture as a replacement for fisheries are contested (Friend and Blake 2009). The option of not building any dams on the lower mainstream has also been suggested by civil society actors and some donors (MRC Meeting 09/08). The language of benefit-sharing has also entered the debate. Benefit-sharing involves developing water resources in their optimal locations and sharing the benefits of this, rather than the water, across the basin (Alam et al. 2009). Benefit-sharing is a relatively new concept, and to date most of what is labelled benefit-sharing either resembles IWRM or could be termed idealistic appeals (Phillips et al. 2006). In the lower Mekong context, benefit-sharing has been evoked in a rudimentary fashion. For example, the economic benefits of hydropower could be shared in terms of sharing the development benefits and foreign exchange revenues earned by the state with the wider community (Bird 2008a). These possible strategies illustrate that mitigation is a difficult and contested terrain.

4.3.3   The MRC and Trade-offs: Opportunities and Challenges

The MRC has recognised, both in its *Strategic Plan 2006–2010* and through the work of the BDP2, that trade-offs are likely as development increases. Identification and negotiation of trade-offs requires knowledge. Knowledge generation and dissemination is one of the ways in which the MRC has entered the debate and is seeking to inform and influence decision making. The MRC recognises that studies and research will have no impact unless they are utilised in decision making (MRC 2006). There are a number of opportunities and challenges associated with this role mainly surrounding how to ensure the impact of the MRC. This is also a contentious role for the MRC as it may involve delivering difficult messages to powerful actors, such as it is not possible to technologically mitigate the fisheries impacts of lower mainstream hydropower dams.

A range of MRC programmes are generating relevant knowledge. These include the Fisheries Programme and BDP2. Scenarios developed by BDP2 will outline the expected changes in the basin under different levels of development. BDP2 hopes these will form the basis of debate about development and 'guide member country development and predict impacts and trade-offs mainly in the hydropower, irrigation and fisheries sectors' (MRC 2008a). The MRC is also conducting activities such as an SEA on mainstream dams. The MRC is situated to bring the transboundary and basin perspectives into discussions and decision making, as its purview includes basin-wide projects, projects which are transboundary or have transboundary impacts, and national projects with significant cumulative basin-wide implications. This role is extremely important as uncoordinated development poses a risk to the basin (Interview Donor State Official 02/08).

The contested terrain of fisheries migration and mitigation illustrates a number of key issues about knowledge and decision making. Different actors have different positions on this issue. The results of the independent Fisheries Expert Group, as mentioned above, were quite conclusive, but Mekong government representatives requested further studies. The MRC argues that it has enough knowledge to engage in the debate, but there are still some important areas requiring more research, including the area of fisheries valuation (Interview MRC Official 06/08). The CEO of the MRC has argued that one of the most important but as yet unanswered questions is the extent to which important fish migrations can be maintained (Bird 2008b). This debate over whether there is enough knowledge about fisheries is related to wider debates about the role of knowledge in decision making.

Donors have questioned whether projects will go ahead if the impacts on fisheries cannot be mitigated (MRC Meeting 09/08). This position is premised on certain assumptions, including that decisions in the Mekong will be objectively based solely on the information and knowledge generated about key questions such as fisheries, and that economic, social and environmental outcomes are all valued equally by decision makers. However, different actors have different interests and value outcomes differently.

Hydropower development is approached differently by various actors. State officials invoke the national interest and approach development from the national level: they argue that poverty levels in the basin mean that hydropower development is necessary, and whilst some people will be impacted, overall the state needs to develop (Interview State Official 06/08). In contrast, civil society actors approach the issues from a local level, or 'bottom-up' approach, arguing that developments should not leave affected people worse off than they were before (Interview Environmental NGO Representative 05/08). Maintaining fisheries is not necessarily seen as development or progress by lower Mekong governments, especially as there are issues surrounding how to quantifiably value fisheries in a way that allows direct comparison with hydropower (Interview Mekong Fisheries Scientist 06/08). These positions are potentially in conflict.

The MRC through knowledge generation and dissemination can bring different perspectives together, but different mindsets and power asymmetries are a challenge that needs to be navigated. Panellists at the 2008 *Mekong mainstream dams: People's voices across borders* conference in Bangkok, Thailand, suggested that the MRC proactively utilise its knowledge to inform debate in a balanced way and challenge the more dominant voices (Lee and Scurrah 2009). Through using its knowledge in this way, the MRC could contribute to mediating power relations between actors and bring different perspectives into debates and decision making. Producing good quality knowledge on relevant issues in an easy to understand, self-explanatory format will give MRC a 'place at the table' in debates (Interview MRC Official 07/08). Disseminating this knowledge to a wide range of actors will provide a common basis for debate. Through this, the MRC can help to create a common understanding of development needs in the basin (Bird 2008b). However, the MRC as an intergovernmental organisation does not have the authority to approve or reject any development projects or act over the wishes of its Member States. This raises questions surrounding how the MRC can ensure impact in the area of trade-offs and navigate challenges such as asymmetric power relations between actors.

How to make the MRC more effective and increase its impact in planning and decision making is one of the key issues the MRC is facing and is trying to navigate (Interview MRC Official 07/08). This is also an important issue for MRC donors. There is a concern that whilst donor funding has helped the MRC to produce a lot of data, knowledge and guidelines, this has not translated into a proactive engagement with Mekong water governance or been utilised to inform decision making (Hirsch et al. 2006; Lee and Scurrah 2009). The intensification of MRC activities since 2008 is an encouraging sign that the organisation is becoming more proactive and engaging with key debates in the region. Translating this into impact in terms of informing, planning and decision making is a key challenge.

There have been some encouraging signs of Member State commitment to the MRC. In 2008, Lao PDR government representatives publicly stated that all Mekong mainstream hydropower development would strictly adhere to the 1995 Mekong Agreement and its procedural rules (MRC 2008b). In June 2008, the Lao government disseminated preliminary information on eight proposed mainstream dams including two joint projects on the Lao-Thai border. The MRC viewed this as a precursor to formal notification and prior consultation (Bird 2008b). However, it remains to be seen if these commitments will be translated into facts on the ground.

In order to increase its impact, the MRC needs to engage more effectively with national planning processes. Commentators are concerned that insufficient linkages between the MRC and national planning processes will impact the MRC's ability to inform decision making (Lee and Scurrah 2009). NMCs provide the formal link between the MRC and the national level. They are largely located in water agencies or environmental departments and ministries, for example, the Lao NMC is part of the Water Resources and Environment Agency. However, this is not necessarily where decision-making power is located. NMCs do not have a high profile in the Member States, and the relationship between MRC activities and national planning is not well established, which has resulted in limited involvement of key line agencies in MRC work activities, and hampered national implementation of MRC initiatives (MRC 2007). The MRC recognises the need to increase engagement with decision makers and line agencies such as the ministries of Finance (Interview MRC Official 07/08). It also needs to make other actors, such as energy ministries and developers, aware of the MRC, its role and the requirements of the 1995 Mekong Agreement (Bird 2008b).

The MRC has increased direct engagement with relevant line agencies since 2008, facilitating dialogues to enhance awareness of the MRC's role. In August 2008, the Lao Department of Energy (located in the Ministry of Energy and Mines) consulted the LNMC and the MRC concerning its studies on the proposed mainstream hydropower development. The Department requested advice and support to ensure that the provisions of the Mekong Agreement were met, that issues such as navigation and fisheries were fully considered and that projects were optimised in an integrated basin context (MRC 2009c). This suggests that the MRC and line agencies are taking initial steps to effectively engage each other, and it is hoped that the MRC can translate this into impact on planning and decision making.

4.4 Public Participation

Public participation is integral to IWRM. It is the subject of the second Dublin Principle which argues that users, planners and policy makers at all levels should be involved in water development and management (GWP 2000). As stated above, different actors have different positions and interests. Stakeholder participation, consultation and dialogue are ways to mediate competing interests. They also allow a range of views to be expressed, and ideally, allow stakeholders to have an input in decision-making processes. The MRC has embraced public participation, especially in terms of facilitating dialogues. Its ability to engage with a range of stakeholders is 'critical if it wants to influence decisions and remain relevant' (Lee and Scurrah 2009: 48). The MRC is actively working to improve and extend its public participation processes, but in order to do so, it must confront a number of opportunities and challenges.

4.4.1   The MRC and Public Participation: An Overview

Public participation in a broad sense encompasses everything from the dissemination of information through to participation in decision-making processes. Stakeholder participation 'through close communication and collaboration with civil society, NGOs and emerging River Basin Organisations' is one of the MRC's guiding management principles and approaches (MRC 2006: ix). At the 2008 MRC *Regional Meeting on Stakeholder Engagement*, the organisation reconfirmed this commitment to public participation, stating that it wanted to mainstream participation in all activities (MRCS 2008). The MRC's *Strategic Plan 2006–2010* roughly defines stakeholders as representing 'both people who have direct interest in the Mekong's water resources as well as people who possess a rich supply of knowledge and opinions to guide planning processes' (MRC 2006: 42). By extending and improving its engagement with stakeholders, the MRC can contribute to the collective pursuit of IWRM, and also increase its relevance to the region's development debates.

The MRC adopted stakeholder principles in 1999. However, initial efforts to engage stakeholders were limited and largely privileged certain types of actors: Member State governments, NMCs and donors. The 2006 Organisational Review argued that Member State governments were seen as the primary, if not only, stakeholder who should be engaged in the MRC (MRC 2007). Engagement with other stakeholders, including civil society, has been limited and through formal processes. In 2001, regional organisations were invited as observers to the MRC's annual governance meetings, including the Council, and Donor Consultative Group meetings. Observers included WWF and IUCN, who also signed MOUs with the MRC. Prior to 2005, observers were able to make a short presentation to the meeting. The reasons why this ceased are unclear (Interview Environmental NGO Representative 06/08). Aside from this process which linked the MRC Council and JC with civil society actors and other development partners, stakeholder engagement with the MRC has largely been conducted through the MRC Secretariat.

The MRC's stakeholder participation has suffered from limited overall strategic direction (MRCS 2008). Some MRC programmes, such as BDP2, have been actively engaging stakeholders in their activities, but an overall organisational framework has been missing. In late 2008, the MRC intensified its efforts to improve and extend its stakeholder participation processes and mechanisms.

Significant activities included a 2008 scoping study on the principles and mechanisms that could guide MRC-wide stakeholder engagement, a 2008 regional meeting on stakeholder engagement and regional stakeholder consultations such as the 2008 *Regional Multi-Stakeholder Consultation on the MRC Hydropower Programme* (hereafter *Hydropower Consultation*). Preliminary ideas for a stakeholder consultative process as part of the annual JC and Council meetings have also been developed. These activities suggest a new openness and willingness on the part of the MRC to engage with a range of stakeholders.

Increased commitment to public participation has arisen for a number of reasons. Stakeholder participation and communication were identified as a weak area by the Organisational Review (MRC 2007). Donors have actively championed public participation within the MRC and its programmes. Initially, public participation was a requirement of the funding agreement between BDP2 and its donors, including Denmark and Sweden (Interview Donor State Representative 05/08). However, BDP2 has a strong commitment to public participation which goes beyond this requirement: public participation permeates the programme. BDP2 is currently engaging a wide range of stakeholders including line agencies, civil society organisations and the research, academic and scientific community in its activities.

Member States have become more open to engagement with different stakeholders and to dialogue on difficult issues. NMC representatives cite the experience of the Cambodian and Vietnamese NMCs meeting with Cambodian NGOs to discuss hydropower development in the 3S Basin as an example of how direct dialogue increases understanding of actors' positions and develops a common understanding (MRC Meeting 05/08). In light of this, some NMC representatives have indicated they would like a direct dialogue over projects such as Don Sahong (MRC Meeting 05/08; Interview State Official 06/08). This illustrates how engaging in public participation activities can lead to more engagement as more powerful actors become more comfortable with participation and hearing alternative views.

In operational terms, facilitating dialogue, consultations and meetings between different stakeholders is one important mechanism through which the MRC can pursue IWRM. Although it is important to remember that public participation is more than meetings and consultations, the MRC is well placed to contribute in this area. It can bring the basin, sub-basin and transboundary perspectives into discussions and engage with and bring together a wide range of stakeholders in the same forum. The MRC argues that the 2008 *Hydropower Consultation* demonstrated that it can act as a facilitator of dialogue at various levels 'not only with governments, but in a multi-stakeholder setting, and with private sector or civil society groups separately on specific issues' (MRC 2008b: 5).

The MRC aims to facilitate dialogue in a number of arenas and has identified a range of stakeholders. This includes ministerial briefings; facilitating dialogue with civil society, line agencies, private sector developers and financiers; and continuing dialogue with China and Burma. The ISH is working to establish a '*representative process with civil society* permitting community views and opinions on hydropower development...to be expressed and discussed' (emphasis in original, MRC 2009c: 29). MRC programmes such as BDP2 have identified affected people, indirectly affected people, donors, the

research, academic and scientific community and international organisations as stakeholders (MRC-BDP2 2009). The MRC is seeking to engage these stakeholders in its
activities.

4.4.2   The MRC and Public Participation: Opportunities and Challenges

The MRC is facing a number of opportunities and challenges in terms of public
participation which it will have to navigate successfully if it wants to contribute
effectively in this area of IWRM. These include questions of openness and access,
privileging certain types of stakeholder, the wider context of civil society engagement in the basin and the question of engagement with local communities.

4.4.2.1   Openness, Transparency and Access to Information

Openness, transparency and access to information are critical for meaningful
stakeholder engagement. The MRC has been criticised in these areas. Some of
these concerns are related to the mandate of the organisation. The MRC is not
accountable to the public but to its Member States. Civil society stakeholders have
called on the MRC to do a number of things which the organisation argues are not
within its mandate, including releasing its review of the EIA for the proposed Don
Sahong hydropower project in Lao PDR (TERRA 2007; MRC Meeting 09/08). The
MRC argues that it cannot release this review as it was commissioned by the
Government of Lao who have not authorised the MRC to release it (Bird 2008a).
Clarifying the role of the MRC in areas such as these will help to improve trust and
confidence between the MRC and various stakeholders.

The MRC has been criticised for having a restrictive approach to releasing or
communicating sensitive information. The 2006 Organisational Review argued that
the MRC's restrictive communication and information disclosure policies threaten
the credibility of the organisation as civil society and scientific organisations
view the MRC as unwilling to release sensitive information which may reveal the
negative consequences of development (MRC 2007). Civil society actors have
repeatedly called for the BDP1 (2001–2006) scenario work to be released. Civil
society representatives believe that the MRC will not release this information
because it is sensitive, and have subsequently labelled the organisation secretive
(Interview Environmental NGO Representative 06/08). The refusal to release
the Don Sahong EIA review has been viewed as another example of the MRC with-
holding information. In response to these criticisms, the MRC developed a new
Communication Strategy and Disclosure Policy which was approved in 2009. It is
currently too early to comment on implementation, but it is hoped that this policy
will mark a new phase of open and clear communication.

The MRC is attempting to increase and create new channels of communication with different stakeholders through mechanisms such as a comments and submissions section on its website. Stakeholders can leave comments and make contributions relevant to the work of the BDP2 and the ISH. Submissions are reviewed by MRC Programme staff and can be posted on the MRC website to help facilitate debate. Submissions through the website are made in English, although submission to the BDP2 can be made in a riparian language via e-mail. The MRC is increasingly utilising its website to disseminate reports, meeting proceedings and presentations. To a certain extent, these mechanisms privilege particular stakeholders, i.e. ones who are computer literate, can read English and have access to the internet. This is not necessarily a constraint if this mechanism is part of a package of mechanisms which targets different stakeholders.

4.4.2.2   Stakeholder Engagement

The MRC has varied engagement with stakeholders. Certain stakeholders, such as the NMCs and donors, have been effectively and extensively engaged. There has been some engagement with civil society through a number of international and regional organisations, networks and NGOs. Stronger engagement with actors such as the private sector, civil society and affected communities is necessary if the MRC is to pursue meaningful participation.

The MRC has successfully engaged Member States government representatives and NMCs, donors and some parts of the research, academic and scientific community. In terms of the hydropower-fisheries trade-off debate, described above, the MRC has facilitated the involvement of the research, academic and scientific community in activities such as the independent Fisheries Expert Group Meeting. The Group's findings were disseminated at the 2008 MRC *Hydropower Consultation* and also through the print and online editions of *Catch and Culture*, the MRC's Fisheries Programme newsletter. Through this process, the MRC brought together scientific and academic expertise and facilitated dialogue between that expertise and other stakeholders. However, as illustrated above, there are some concerns over how willing government stakeholders are to accept the results of this expertise and use it within decision making.

The MRC has extensively engaged donors in the hydropower-fisheries debate. Donors have raised issues both publicly in forums such as the 2008 MRC *Hydropower Consultation* and in their private meetings with the MRC. However, there has been some misunderstanding between donors and Member State government representatives. Due to the possible impacts of extensive hydropower development, some donors have advised caution (MRC Meeting 09/08; Interview Donor State Official 05/08). Some government stakeholders have interpreted this as an anti-hydropower stance. This has affected the openness and frankness of some of the dialogue between donors and Member governments (Interview Donor State Official 05/08). This illustrates how a key public participation challenge is generating trust and confidence amongst different actors so that they can participate openly and honestly. Through facilitating dialogue, the MRC can assist different stakeholders to understand each other and to communicate clearly.

The MRC needs to improve its engagement with certain stakeholders. This includes the private sector and developers who are key actors driving hydropower development in the region. These actors are largely unaware of the MRC, its work and requirements under the 1995 Mekong Agreement (Bird 2008b). The MRC is attempting to address this issue. In 2008, the MRC organised a Developers Workshop hosted by the Department of Energy, Lao PDR. The MRC's roles and responsibilities under the 1995 Mekong Agreement and its capabilities in various areas relevant to hydropower development were presented. It was also an opportunity for participants, including developers, to exchange information on proposed projects (MRC 2009c). Through more active engagement with the private sector, the MRC can bring the basin perspective into discussions.

4.4.2.3   Power, Participation and Civil Society

Questions of power and access are extremely important in terms of public participation and facilitating dialogue. It is vital that stakeholders from all levels are involved in debate and decision making over hydropower development in the lower Mekong, especially local communities who are likely to be most adversely affected by any large-scale infrastructure development. Engagement of civil society and local communities raises the important question of power asymmetries. All actors are not equal in terms of the power they wield in society and decision making. This raises the concern that patterns of participation can reflect power asymmetries rather than even them out (Molle 2008). Consequently, the question becomes how to produce meaningful participation processes that mediate these differences in power. In order to successfully facilitate dialogue and engage a range of stakeholders, the MRC needs to address these questions.

Public participation is connected to and affected by the political dynamics in which it takes place. The MRC context is further complicated by the basin's transboundary nature and by the extent to which Member States allow participation internally (Sneddon and Fox 2007). Hirsch et al. (2006) argue that until the MRC can understand how the interests of stakeholders are accommodated and mediated at the national and local level and engage with these stakeholders directly, it is 'hard to see how it can be truly effective as an agent' of IWRM (xviii). The lower Mekong basin comprises four states, and public participation differs from state to state (MRC Meeting 05/08). The space for civil society participation in society and in decision making is contested in the lower Mekong basin. To address questions of power asymmetries, the MRC will need to understand and find ways to operate within this dynamic.

Civil society in the lower Mekong is limited, although it does vary from state to state, as do spaces for action. A robust social and environmental movement exists in Thailand, uniting domestic NGOs, communities, activists, academics and international NGOs. This movement has challenged hydropower projects such as Pak

Mun dam in Thailand and organised conferences such as the 2008 *Mekong mainstream dams: People's voices across* borders, which brought together civil society, government and international stakeholders. Civil society is growing in Cambodia with the emergence of NGOs that combine domestic and international elements. These include the Fisheries Action Coalition Team (active on fisheries issues in the Tonle Sap and Cambodian provinces in the Mekong basin) and the 3S Protection Network (active on hydropower issues in the 3S basin and assisting dam affected communities). International NGOs including WWF are active in Lao PDR and Vietnam, but domestic civil society is largely streamlined through state representatives. For example, the Lao Women's Union is a mass organisation formed in 1955 by the Central Committee of the Lao People's Revolutionary Committee.

Differences in both civil society spaces for action and what constitutes civil society affect its ability to engage with and participate in development debates and decisions. Hirsch (2001) argues that there is limited political space in the Mekong states to 'articulate concerns over projects and other aspects of development that threaten social and environmental sustainability' (245). Concern has also been expressed that affected communities are not able to openly express their views when consulted about projects. For example, concern has been expressed that in initial local consultations for the Nam Theun 2 hydroelectric project, Lao PDR, local stakeholders had limited ability to express their opinions without fear of reprisals and tended to agree with government representatives (Chamberlain 2004). This illustrates how the wider political environment impacts public participation and also that hosting a consultation or a dialogue is not necessarily congruent with meaningful public participation. The MRC will need to understand and navigate the differences in civil society if it is to improve its engagement with a range of stakeholders.

Elements of international and regional civil society have been well engaged by the MRC in its activities. This includes joint activities with international NGOs such as WWF; the involvement of research, academic and scientific organisations such as MPOWER in the work of the BDP2; and participation in MRC consultations by NGOs such as International Rivers. However, links with regional, national and local civil society groups need to be strengthened. There are a small number of lower Mekong NGOs and civil society networks such as Living River Siam and the Rivers Coalition in Cambodia who attend MRC consultations and feel comfortable to participate in terms of making presentations, asking questions of the MRC and government representatives and entering into debate. These engagements with civil society are a positive sign of the MRC's commitment to public participation, but in order to improve its public participation processes, the MRC should involve a wider range of regional, national and local civil society organisations.

The MRC as a basin organisation can bring civil society representatives and perspectives into its discussions with planners and decision makers. MRC consultations represent one sphere in which civil society representatives can 'get their voices heard' by planners and decision makers, especially in an environment where domestic opportunities to do so may be limited. Commentators have welcomed proposals to formalise a consultative process for stakeholders as part of the MRC's annual governance meetings. This would allow civil society and other stakeholders to convey their concerns directly to the MRC's governing bodies and engage with ministers and line agencies (Lee and Scurrah 2009). By providing opportunities for different stakeholders to interact, and by facilitating dialogue, the MRC can introduce different perspectives as well as increase trust and confidence between stakeholders. Trust between stakeholders has been a problem in the region. However, consultations and dialogues have shown certain actors that participation is not an opportunity for other actors to criticise them. Subsequently, they have contributed to an enabling environment where different points of view can be shared (Interview Environmental NGO Representative 06/08). This illustrates that the MRC has the potential for a strong role in terms of providing spaces for interaction and increasing trust.

One of the key criticisms of the MRC's stakeholder engagement is that affected communities and other local stakeholders are not involved. Participants at the 2008 *Hydropower Consultation* questioned whether the consultation was representative when local stakeholders and community representatives were absent (MRC 2008b). The absence of those most affected by proposed developments – farmers, fishermen and local communities – was raised a number of times during the *Hydropower Consultation*. Whilst civil society has been engaged through NGOs, some civil society representatives argue that this is not enough: affected communities have to be involved also (MRC Meeting 09/08). Engaging effectively with civil society and local communities is one of the key challenges that the MRC must navigate in order to effectively engage in public participation.

The MRC identifies those affected by a project as a stakeholder, arguing that they should be involved in the planning, implementation and monitoring of the project (MRC 2006). The MRC has recognised that it needs to do more to engage communities and ensure their views are reflected (MRC 2008b; MRC Meeting 09/08). Large consultations are not the mechanism for this engagement. Instead, the MRC argues it needs to access the sub-area and local level, and could do so through the work of BDP2 whose sub-area work divides the basin into ten sub-areas (MRC 2008b; MRC Meeting 09/08), although it is still unclear how local communities will be engaged in this. Large-scale, high-profile consultations could be intimidating to local level stakeholders, resulting in them not being able to express their views. Within larger, regional-level meetings, the MRC is dependent on civil society organisations to represent communities' views. Consequently, it is important that the MRC strengthens its civil society engagement at the same time as it explores ways to engage local communities.

4.5   Conclusion

The MRC is committed to implementing IWRM in a development context with a strong infrastructure focus. There are a multitude of plans for hydropower development on the lower Mekong River supported by justifications which link hydropower, socio-economic development and poverty reduction. However, hydropower development will have a number of social and environmental impacts.

Within this development context, the MRC is committed to implementing IWRM. IWRM comprises three goals: economic efficiency, social equity and environmental sustainability. In the Mekong context, there are a number of potential tensions between these goals, and how to balance them and ensure equitable development is an unfolding challenge. The MRC has the potential to play a strong role in the collective pursuit of IWRM in two key areas: the identification and negotiation of trade-offs and public participation. However, the MRC is facing a number of challenges which it must navigate successfully in order to fulfil this potential.

Trade-offs are extremely contentious in the lower Mekong. They are being presented and debated in terms of economic considerations on the one side and social and environmental on the other. Hydropower development will negatively impact on fish migration, and it is unclear whether this can be mitigated either through technological mitigation such as fish passages or alternative means such as replacement livelihoods. The MRC can place a strong role in terms of knowledge generation and bringing the basin, transboundary and alternative perspectives into the debate.

There are challenges associated with the MRC's knowledge role. Generating knowledge is not enough. The MRC has to proactively confront issues and ensure that its knowledge has an impact in terms of informing debates and decision making. In order to do this, the MRC is increasing its engagement with decision makers. It is also engaging in a number of activities, such as a SEA of the lower Mekong mainstream dams, which will increase the levels of understanding about the cumulative impacts of the proposed hydropower projects. These are encouraging signs on the part of the MRC, although it must be remembered that the organisation's mandate means that it cannot impose its findings on Member States.

The MRC also has a strong potential to play a role in IWRM in terms of public participation. Despite criticisms of the MRC's early attempts, it is now seeking to improve and extend participation mechanisms. The MRC as a basin-level organisation is able to bring a wide range of stakeholders together including government stakeholders, donors and civil society. Previously, certain stakeholders such as Member State governments have been privileged in terms of access and information. However, the MRC is now consolidating and strengthening its relations with civil society actors. This is especially important in a context where domestic space for civil society may be limited. The MRC and its public participation processes represent a way in which civil society stakeholders are able to engage with decision makers when otherwise this may not be possible.

Questions of power and access are extremely important. If the MRC is to effectively engage in public participation, it must confront and navigate these difficult issues. The MRC is currently extending the range of stakeholders it engages in order to bring different perspectives together. Through facilitating dialogues, the MRC can contribute to building trust between different stakeholders, allowing them to discuss difficult issues. Through providing knowledge and neutral spaces the MRC can help to overcome some of the challenges associated with asymmetric power relations between stakeholders. One of the key challenges facing the MRC is how to engage local communities and ensure that the voice of those most affected by developments is heard.

The MRC has a strong potential to pursue IWRM in terms of trade-offs and public participation. However, there are both opportunities and challenges which it must navigate if the organisation is to realise its potential. The MRC is currently proactively facing a number of these.

References

ADB and World Bank (2006) WB/ADB joint working paper on future directions for water resources management in the Mekong river basin: Mekong water resources assistance strategy (MWRAS). World Bank, Vientiane

Alam U, Dione O, Jeffrey P (2009) The benefit-sharing principle: implementing sovereignty bargains on water. Political Geogr 28(2):90–100

Backer EB (2007) The Mekong river commission: does it work, and how does the Mekong basin's geography influence its effectiveness? Sudostasien aktuell 4:31–55

Bird J (2008a) Requests from Mekong civil society. Letter sent to TERRA, 11 Apr 2008. http://www.terraper.org/articles/MRCS%20CEO%20Response%20to%20TERRA.pdf. Accessed 12 July 2008

Bird J (2008b) Integrated water resources management in a rapidly-growing private sector development context: the Mekong basin. Presentation at the world water week, Stockholm Asia day: theme 5 – promoting effective basin management, 19 Aug 2008. http://www.mrcmekong.org/MRC_news/CEO%20MRCS%20speech%20WWW%20Stockholm%2008.pdf. Accessed 12 Dec 2008

Bird J (2008c) Hydropower in the context of basin-wide water resources planning. Presentation at MRC regional multi-stakeholder consultation on the MRC hydropower programme. http://www.mrcmekong.org/download/Presentations/regional-hydro/1-1%20Integration%20-%20CEO%2025sep08.pdf. Accessed 14 Dec 2008

Biswas A (2004) Integrated water resources management: a reassessment, a water forum contribution. Water Int 29(2):248–256

Chamberlain J (2004) Proposed Nam Theun 2 hydroelectric project assessing the quality of local consultations. World Bank, Vientiane

Coates D, Ouch P, Ubolratana S, Tung NT, Sinthavong V (2003) Biodiversity and fisheries in the lower Mekong basin. Mekong development series no. 2. Mekong River Commission, Phnom Penh

Dugan P (2008) Mainstream dams as barriers to fish migration: international learning and implications for the Mekong. Catch Cult 14(3):9–15

Friend R, Blake D (2009) Negotiating trade-offs in water resources development in the Mekong basin: implications for fisheries and fishery-based livelihoods. Water Policy 11(Supplement 1):13–30

Global Water Partnership (2000) TAC background paper no.4: integrated water resources management. Global Water Partnership, Denmark

Government of Lao PDR (2003) National growth and poverty eradication strategy (NGPES). Government of Lao PDR, Vientiane

Hang PT, Lennaerts T (2008) Joint basin development planning to foster IWRM in the lower Mekong basin. Paper presented at Mekong management at a watershed- IWRM in the global crisis? Symposium, Gothenburg University, Sweden

Hirsch P (2001) Globalisation, regionalisation and local voices: the Asian development bank and re-scaled politics of environment in the Mekong region. Singap J Trop Geogr 22(3):237–251

Hirsch P (2006) The Mekong river commission and the question of national interest(s). Watershed 12(1):20–25

Hirsch P, Jensen KM, Boer B, Carrard N, FitzGerald S, Lyster R (2006) National interests and transboundary water governance in the Mekong. University of Sydney/Australian Mekong Resource Centre, Australia

Hortle KG (2007) Consumption and the yield of fish and other aquatic animals from the lower Mekong basin. MRC technical paper no 16. Mekong River Commission, Vientiane, Lao PDR

International Rivers (2008) Power surge: the impacts of rapid dam development in Laos. International Rivers, Berkley

Interview (02/08) Donor state official. Vientiane, Lao PDR

Interview (05/08) Donor state official. Vientiane, Lao PDR

Interview (05/08) Environmental NGO representative. Bangkok, Thailand

Interview (06/08) Mekong fisheries scientist. Vientiane, Lao PDR

Interview (06/08) MRC official. Vientiane, Lao PDR

Interview (06/08) Environmental NGO representative. Vientiane, Lao PDR

Interview (06/08) State official. Vientiane, Lao PDR

Interview (07/08) MRC official. Vientiane, Lao PDR

Jonch-Clausen T (2004) "…Integrated water resources management (IWRM) and water efficiency plans by 2005" why, what and how? Global Water Partnership, Denmark

King P, Bird J, Hass L (2007) Environmental criteria for hydropower development in the Mekong region. WWF, Vientiane

Klopper Y (2008) Southeast Asian water conflicts from a political geography perspective. Asia Eur J 6(2):325–343

Lang MT (2006) Management of the Mekong river basin: contesting its sustainability from a communication perspective. In: Tvedt T, Jakobsson E (eds) A history of water: water control and river biographies. I. B. Tauris, London, pp 552–580

Lee G, Scurrah N (2009) Power and responsibility: the Mekong river commission and lower Mekong mainstream dams. The University of Sydney, Sydney

Mekong Agreement (1995) Agreement on the cooperation for the sustainable development of the Mekong river basin. Mekong River Commission, Chiang Rai

Middleton C (2008) The sleeping dragon awakes: China's growing role in the business and politics of hydropower development in the Mekong region. Watershed 12(3):51–64

Molle F (2008) Nirvana concepts, narratives and policy models: insights from the water sector. Water Altern 1(1):131–156

Molle F (2009) River-basin planning and management: the social life of a concept. Geoforum 40(3):484–494. doi:10.1016/j.geoforum.2009.03.004

MRC (Mekong River Commission) (2003) State of the basin report. Mekong River Commission, Phnom Penh

MRC (Mekong River Commission) (2005) Strategic directions for IWRM in the lower Mekong basin. Mekong River Commission, Vientiane

MRC (Mekong River Commission) (2006) Strategic plan 2006–2010. Mekong River Commission, Vientiane

MRC (Mekong River Commission) (2007) Independent organisational, financial and institutional review of the Mekong river commission secretariat and the national Mekong committees. Mekong River Commission, Vientiane

MRC (Mekong River Commission) (2008a) Stakeholder consultation on MRC's basin development plan phase 2 (BDP 2) and its inception report. Mekong River Commission, Vientiane

MRC (Mekong River Commission) (2008b) MRC regional multi-stakeholder consultation on the MRC hydropower programme. Mekong River Commission, Vientiane

MRC (Mekong River Commission) (2009a) Mid-term review of the Mekong river commission strategic plan 2006–2010, final report, Jan 2009. Mekong River Commission, Vientiane, Lao PDR

MRC (Mekong River Commission) (2009b) Global economic crisis provides opportunities for sustainable development says Mekong body. Press release. http://www.mrcmekong.org/MRC_news/press09/29th_JC_27-mar-09.htm. Accessed 15 Apr 2009

MRC (Mekong River Commission) (2009c) Initiative on sustainable hydropower: work plan draft 1. Mekong River Commission, Vientiane

MRC (Mekong River Commission) (2009d) IWRM-based basin development strategy for the lower Mekong basin, Incomplete consultation draft no 1, Oct 2009. Mekong River Commission, Vientiane, Lao PDR

MRC-BDP2 (2009) Stakeholder participation and communication plan for basin development planning in the lower Mekong basin. Mekong River Commission, Vientiane

MRCS (Mekong River Commission Secretariat) (2008) Summary report of the MRC regional meeting on stakeholder engagement. Mekong River Commission, Vientiane

Participant Observation Notes (05/08) MRC meeting. Vientiane, Lao PDR

Participant Observation Notes (09/08) MRC meeting. Vientiane, Lao PDR

Participant Observation Notes (11/07) MRC meeting. Siem Reap, Cambodia

Phillips DJH, Daoudy M, Öjendal J, Turton A, McCaffrey S (2006) Trans-boundary water cooperation as a tool for conflict prevention and for broader benefit-sharing. Ministry for Foreign Affairs, Stockholm

Poulsen AF, Hortle KG, Valbo-Jorgensen J, Chan S, Chhuon CK, Viravong S, Bouakhamvongsa K, Suntornratana U, Yoorong N, Nguyen TT, Tran BQ (2004) Distribution and ecology of some important riverine fish species of the Mekong river basin. MRC technical paper no 10. Mekong River Commission, Phnom Penh, Cambodia

Sneddon C, Fox C (2006) Rethinking transboundary waters: a critical hydropolitics of the Mekong basin. Political Geogr 25(2):181–202

Sneddon C, Fox C (2007) Power, development, and institutional change: participatory governance in the lower Mekong basin. World Dev 35(12):2161–2181

TERRA (2007) MRC silent as mainstream dams move forward. Press briefing. http://www.terraper.org/articles/Mekong%20Mainstream%20dams%20media%20brief%208Nov07.pdf. Accessed 15 Dec 2007

Varis O, Rahaman M, Stucki V (2008) The rocky road from integrated plans to implementation: lessons learned from the Mekong and Senegal river basins. Int J Water Resour Dev 24(1):103–121

World Bank (2004) Modeled observations on development scenarios in the lower Mekong basin. World Bank, Vientiane

Chapter 5
Domestic Water Policy Implications on International Transboundary Water Development: A Case Study of Thailand

Naho Mirumachi

Abstract The purpose of this chapter is to understand the impact of domestic water policies on international transboundary water development and management. Transboundary water allocation and river development are part of a political process in which different interests of basin states are reflected. The chapter posits that the national hydrocracy can execute control over the promotion or demotion of water allocation rules and policies, and water development plans in transboundary river basins. The chapter analyzes Thailand's political economy of water and how the domestic water development progress has manifested in the formation and operation of the regional water management institutions: the Mekong Committee, the Interim Mekong Committee, and the Mekong River Commission. The chapter uses the concepts of hydraulic mission and reflexive modernity to analyze water management paradigms. By showing the progress of water development and the supporting water policies, the chapter examines the concerns of the Thai hydrocracy vis-à-vis regional water management. Specifically, it will be shown how the Thai hydrocracy politicized and securitized issues of water allocation and utilization based on their water development plans and concerns of institutional rules. Some policy implications regarding the way domestic policies can impact transboundary water management, especially transboundary IWRM, are discussed.

5.1 Introduction

The politics of water management are interconnected at many levels. Mollinga (2008) identified various "domains" of politics: global politics of water discourse; hydropolitics between sovereign states; domestic politics over national water policy; and

N. Mirumachi (✉)
Department of Geography, King's College London, Strand, London WC2R 2LS, UK
e-mail: Naho.mirumachi@kcl.ac.uk

J. Öjendal et al. (eds.), *Politics and Development in a Transboundary Watershed:*
The Case of the Lower Mekong Basin, DOI 10.1007/978-94-007-0476-3_5,
© Springer Science+Business Media B.V. 2012

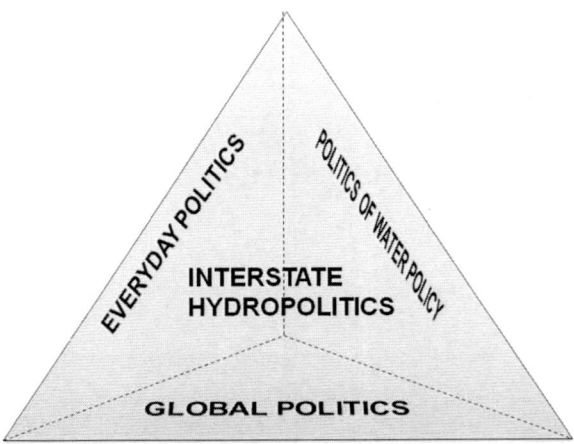

Fig. 5.1 Prism of water politics based on Mollinga's (2008) domains of water politics

"everyday politics" of day-to-day water resources management between individuals and small stakeholder groups. The fact that Mollinga stated water management politics as domains and not as layers implies the interconnected nature of the politics over water. Considering the complexity of water uses, these domains are perhaps not stacked on top of each other from local to global, forming a linear, vertical composition, but rather interconnected in the shape of a prism (Fig. 5.1). If the politics of water is understood as domains interconnected in a prism, the analysis of transboundary river basins can benefit from studies that attempt to understand how the domestic water politics and policies shape the practices and discourse of transboundary water management. In this chapter, the impact of domestic water policies on transboundary water management and development is explored.

For over five decades, the Lower Mekong River basin has been governed by institutions that provide a multilateral platform for river basin development. The Committee for Coordination of Investigations of the Lower Mekong Basin (herein-after Mekong Committee (MC)) was established in 1957 by four countries, Laos, Thailand, Vietnam, and Cambodia, in order to facilitate development of the river. Since then, the countries have maintained an institutional setup for facilitating river development through the Interim Mekong Committee (IMC) and Mekong River Commission (MRC). There have been positive assessments of the past organizations for maintaining a forum for multilateral dialogue and a general spirit of cooperation despite high regional and national insecurity (Browder and Ortolano 2000; Jacobs 1995). However, some analysts have mentioned that national interests of the basin states have hindered the overall effectiveness of the current multilateral framework (Öjendal 2000; Hirsch et al. 2006; Keskinen et al. 2008). These studies point out the difficulty of implementing transboundary measures within a context of diverse riparian interests. In such a context, it is important to understand how water issues are conceptualized and prioritized within the national discourse of the states and identify the reasons why the transboundary measures are left ineffective.

In particular, as the current MRC framework adopts Integrated Water Resources Management (IWRM), which in theory "promotes the co-ordinated development and management of water, land and related resources, in order to maximize the resultant economic and social welfare in an equitable manner without compromising the sustainability of vital ecosystems" (GWP 2000:22), it is pertinent to question the influence of domestic water politics on international hydropolitics. The concept of IWRM encourages coordination between the national and international levels of water resources management, and the success of IWRM would be influenced by the consistencies between the two.

The chapter analyzes the progress of Mekong water development in Thailand. The Mekong River forms the border with Laos in the north and northeast regions of Thailand, forming one of the 25 major river basins in Thailand. Thailand is one of the key states for the development of the basin. Along with Vietnam, Thailand is described as a hegemonic state and has exerted much political power in the negotiations of the multilateral institutions (Makim 2002). Thailand's progress in water development highlights the changes in the political economy of water since the mid-1950s. The chapter analyzes how such changes have caused the hydrocracy of Thailand to politicize and securitize water issues in the negotiations of the multilateral institutions. The hydrocracy is described as a body of stakeholders who pursue progress and development by controlling nature based on technical, scientific means, often comprised of the ministries responsible for agriculture, irrigation, or water resources (Wester 2008). In the case of Thailand, the hydrocracy includes the governmental departments that are involved in the national management of water, such as the Royal Irrigation Department and organizations under the Office of the Prime Minister, and also the ministries that are responsible for international water negotiations, such as the Ministry of Foreign Affairs, former Ministry of Science, Technology and Environment (later reorganized to form the Ministry of Energy and Ministry of Natural Resources). It also includes the political leaders who have taken specific interests in water management. Section 5.2 provides conceptual ground for examining the changes of the political economy of water and impacts on water management at the international level. Section 5.3 examines the process of water resource development in Thailand during the MC, IMC, and MRC periods up to 2006. Section 5.4 analyzes the impact of the Thai water policies on international water resource development within the three institutions. The chapter ends with a discussion on the implications of domestic water policies on transboundary IWRM promoted by the MRC.

5.2 Water Management Paradigms and Hydraulic Control

To analyze the political economy of water and its changes, the evolution of water management principles offers some indications. In a spatial overview of water management paradigms, Allan (2003) argued that limited water use during the premodern times (first paradigm) increases dramatically during industrial modernity (second paradigm). Water management is facilitated by the central authority embarking on the hydraulic mission. The hydraulic mission is a phase of investment in dams, irrigation schemes, and ground water extraction projects organized by the state to capture water. Large-scale hydraulic investments are a key feature. However, once water scarcity and environmental degradation become severe owing to excessive water capture, water management changes to more economical and ecological uses of water. Hooper (2003:14) characterized this change as one caused by the ignorance of "the more diverse range of resource use features of river basins that interact to create the so-called 'wicked' problems of environmental management and sustainable water resources management."[1] Once in this reflexive modernity paradigm, water use, in theory, starts to decline. Environmental concerns reallocate water to the basin for biodiversity and sustainability of the water environment. Economic water measures, such as agricultural efficiency and water reuse, are implemented. IWRM fits into this reflexive paradigm as it takes into account water for human activities and environment, water for the rich and poor, and linkages between water and land.

The paradigms reflect how states value water and its uses. For example, states executing the hydraulic mission will prioritize investment in infrastructure for water capture and for expanding their water capture potential. Taking in the "diverse range of resource use," riparian states in reflexive modernity will call for basin-wide management that promotes sustainable use of water resources, including the harmonization of upstream and downstream development. Allan's conceptualization highlights the dominant discourse that determines water management principles, and it should be expected that plural competing discourses exist.

The main facilitator of the hydraulic mission is the hydrocracy (Wester 2008). The chapter posits that the hydrocracy can execute control over the promotion or demotion of national and international projects in transboundary river basins. Put differently, the chapter posits that the hydrocracy can politicize and securitize the discourse of water resources development at the international level. Politicization and securitization are political processes in which certain issues become highly prioritized on the political agenda of a state and provoke actions to safeguard state interests (Buzan et al. 1998). When issues are politicized, they are items on the national agenda requiring resources to be allocated. Politicization creates a space for "*discussion, debate* and *deliberation* [emphasis in original]" (Roe 2006:426). When issues are securitized, it closes the political space for deliberation such that extraordinary measures out of bounds of normal politics are taken (Buzan et al. 1998). These political processes, as exemplified in securitization theory, can also be applied to environmental issues, especially in water resource development of transboundary river basins where water allocation comes with heavy baggage of sovereignty and

---

[1] Wicked problems were first described by Horst Rittel in his coauthored paper (Rittel and Webber 1973). The paper described a wicked problem as an evasive and unique problem to which many conceivable solutions must be proposed for continual solution (Rittel and Webber 1973).

territorial concerns. Political space is closed down to securitize an issue by framing it as an existential threat (Buzan et al. 1998). For example, water scarcity may be framed as a threat to the agricultural sector. The process of politicizing and securitizing water resource development can be used by the hydrocracy to legitimize its intervention by the construction of threats (Warner 2000). The hydrocracy can argue that there is water scarcity, which may limit economic growth of the state, thereby legitimizing the construction of dams and irrigation canals. Based on the reason that it would be a matter of survival of the state, the hydrocracy can frame the implementation of certain transboundary projects as necessary or harmful, in which case the project would be promoted or blocked from being implemented. Such hydraulic control of the hydrocracy can be examined through the negotiations and implementation of river development projects and water resource management.

5.3   The Thai Hydraulic Mission and Inclusive Water Resource Management

This section examines the progress of water resources development in Thailand by applying the concepts of hydraulic mission and reflexive modernity of Allan's (2003) water management paradigms. The analysis of the political economy of water is explained in the context of the three regional water management institutions spanning from 1957 to 2006: the MC period during 1957 and 1978, the IMC period during 1978 and 1995, and MRC period during 1995 and 2006. The scope of analysis is limited to 2006 when the first phase of transboundary IWRM plans was completed.

5.3.1   Postwar Hydraulic Mission

While irrigation and rice production had long been part of the agricultural tradition of Thailand, there was rapid development in irrigation after the Second World War (Kaida 1978). The global food shortage after the war enhanced the necessity to increase production. Studies by the United Nations Economic Commission for Asia and the Far East (ECAFE) in the 1950s proposed extensively utilizing the Mekong River waters with large-scale dams, effectively promoting the hydraulic mission. In particular, developing a dam on the mainstream that flows in the northeast region of Thailand was given much attention (United Nations Survey Mission 1958).

The northeast region was an area that the central government viewed as strategically important for two reasons. First, rural population was increasing after the end of the war, causing food security concerns. Comparatively, this region was experiencing slow progress in agricultural development. While mean rice yields were high in the provinces along the Mekong River in the northeast region, the rice

Table 5.1 Irrigation projects in northeast Thailand proposed in the Indicative Basin Report (Source: Interim Mekong Committee 1985: Annex 1)

Project name	Year of completion	Irrigation area (ha)	Installed capacity (MW)
Nam Pung	1965	0	6.3
Nam Pong	1966	53,000	25.0
Lam Takong	1970	38,000	0
Lam Dom Noi	1971	24,000	24.0
Lam Pra Plerng	1971	9,760	0
Nam Oon	1973	32,000	0
Nam Phrom	1973	0	40.0
Lam Pao	1975	21,300	0
Pump irrigation on the Mun/Chi stage 1	1982	6,500	0

production per unit of area and rice production per agricultural resident were lagging behind compared to the central plains of the Chao Phraya basin region (Behrman 1968). Second, the central government was concerned with regional security. The northeast region received the central government's attention and investment so as to prevent communist insurgency from across the Lao border and ensure political stability within the region (Donner 1978; Feeny 1982; Chomchai 1994).

Recognizing the importance of developing the northeast region, the Government of Thailand emphasized regional planning when it first published its National Economic and Social Development Plan (1961–1966). In the Second National Economic Development Plan, development of this region was understood "to raise the standard of living as quickly as possible through various development pro-grammes. Scarce natural resources and inferior topographic conditions necessitate such projects as potable water supply, communication and power. Furthermore, subversive activities and infiltration in this part of the country make the policy a matter of top priority" (Government of Thailand 1966: 95–96 in Donner 1978). The northeast region received not only national but also international interest. The USA, keen to suppress communist influence expansion, invested in rural development projects (Chomchai 1994). Irrigation area expansion was achieved during the 1960s, and despite the seasonal rainfall, which causes water scarcity and flooding, the northeast region experienced increasing median mean annual provincial growth rate of rice production (Behrman 1968:161).

Many of the projects in the northeast were incorporated into the MC activities. The Indicative Basin Report published in 1970 by the MC compiled numerous studies assessing potential projects that would harness the hitherto untapped resource of the Mekong. Irrigation expansion and hydropower development was a priority, and both mainstream and tributary projects were designed to achieve these goals. In the report, eight irrigation and hydropower projects on the tributaries were dedicated to the northeast of Thailand (see Table 5.1). These projects provided approximately 17,800 ha of potential irrigation area (IMC 1985). Of these, four dams had a total installed capacity of 95.3 MW (IMC 1985). In addition, the

Indicative Basin Report also stressed that a high dam above Vientiane, the Pa Mong dam, would be beneficial in the long term to meet the irrigation needs of the immediate dam region and the whole lower river basin. The Pa Mong dam would have a full reservoir level of approximately 250 m and provide irrigation, hydropower, and flood control benefits (Hori 1996). Thailand would benefit from increased irrigation area in the northeast and supply of hydropower.

Initiating hydraulic investment was important for Thailand at this stage in time. In other words, agricultural productivity in comparison to the amount of investment was not scrutinized. While dams constructed in the northeast region allowed water to be stored, infrastructure to utilize the stored waters was incomplete. The investments in large irrigation projects in the northeast were achieved because of the regional development policy: "And the justification of these low-efficiency projects seems to rest on making 'water' the basis of regional development in the Northeast" (Kaida 1978:225).

However, the Pa Mong dam and other mainstream projects were not implemented because of regional instability. The Vietnam War was intensifying in the region. The factions within the MC became even more evident between Thailand and the communist-influenced Indo-Chinese countries (Laos, Vietnam, and Cambodia). Diplomatic relations were dissolved between Thailand and Vietnam during 1975–1978, and the function of the MC effectively came to a halt. The defunct commission was revived in 1978 without the participation of Cambodia. Under the Khmer Rouge regime, Cambodia isolated itself from the international community and ceased to send representatives to the multilateral committee. Despite the committee being revived as the Interim Mekong Committee (IMC), the *Joint Declaration of the Principles for Utilization of the Waters of the Mekong Basin* established in 1975 required the consent of all four countries on mainstream development; without the participation of Cambodia, it was impossible to execute plans according to the Indicative Basin Report.

5.3.2 *"Two-Pronged" Hydraulic Mission*

The absence of Cambodia and the growing regional instability meant that much of MC's international activities downscaled, and many of the hydraulic projects after the late 1970s were unilateral, domestic projects. Thailand continued to devote resources to irrigation expansion. Overall, irrigation area increased rapidly in the first decade of the IMC (see Table 5.2). However, the development of the northeast region still lagged behind, and there were concerns about its limit on agricultural productivity. Much land had been exploited for irrigation, and so efficiency of land utilization was a challenge (IMC 1985).

In the early years of the IMC period, the Fourth National Economic and Social Development Plan (1977–1981) of the Government of Thailand specified the Mekong waters as a necessary resource for agricultural production, in particular, in the northeast region: "water projects have to be implemented on a region-wide basis

Table 5.2 Thailand's irrigation growth (Source: Adapted from Budhaka et al. 2002)

National Economic and Social Development Plan	Year	Irrigation area		% Irrigation area over total area	Capacity (million m³)
		(million rai)	(million ha)		
First Plan	1961–1966	9.72	1.58	3.0	14.4
Second Plan	1967–1971	10.96	1.78	3.4	15.0
Third Plan	1972–1976	14.38	2.33	4.4	24.3
Fourth Plan	1977–1981	15.84	2.57	4.9	25.4
Fifth Plan	1982–1986	18.71	3.04	5.8	28.6
Sixth Plan	1987–1991	20.71	3.36	6.4	30.2
Seventh Plan	1992–1996	21.68	3.52	6.7	31.6
Eight Plan	1997–2001	22.39	3.64	6.9	32.3
Ninth Plan[a]	2002–2006	28.49	4.63	8.8	36.5
Tenth Plan[a]	2007–2011	30.71	4.99	9.5	39.2

[a]Expected figures

and more water from the Mekong River must be pumped and channelled into irrigation canals" (Government of Thailand 1976: 155 in Sneddon 2003:2240). However, funding through the IMC became limited (Dixon 1999). In addition, it became gradually clear that the Pa Mong dam would not be implemented (Nakayama 1999; Floch et al. 2007), despite the IMC still remaining positive about the option: "… while it would not be possible to reach decisions on the implementation of the Pa Mong scheme in the absence of the fourth riparian member, further Pa Mong studies remained desirable" (IMC 1983a:1).

It was against this wider political economy of transboundary water development that Thailand developed its "two-pronged water policy" (Floch et al. 2007). The idea was to develop small- and large-scale water supply projects (Floch et al. 2007). The Government of Thailand constructed numerous pump irrigation schemes in the Mun and Chi tributaries of the Mekong, located in the northeast region. The first stage of this pump irrigation scheme achieved 6,500 ha of irrigated area in 1982, and the second scheme was designed to further augment by another 10,000 ha (IMC 1985). While the small-scale projects were providing much-needed water supply to the region, the large-scale projects were effectively the sole option for further increasing irrigation capacity (Floch et al. 2007). As a large-scale project, "Green Isaan" was proposed in 1987. Initiated by the Thai military and later involving the National Economic and Social Development Board and other governmental agencies, the project aimed to relieve water stress in the northeast and accelerate development through irrigation (Floch et al. 2007). "Green Isaan" was not implemented but replaced by the Kong-Chi-Mun (KCM) project. This project was designed to develop floodplain storage and water diversion in the Chi-Mun basin. The Chi-Mun basin, where the Chi River drains into the Mun River, has a basin area of 119,570 km² and is the largest subbasin of the Mekong River (Tingsanchali and Singh 1996). While figures vary, the KCM project would irrigate between 796,800 and 1,277,700 ha and also have a hydropower component of 25-MW capacity

(Tingsanchali and Singh 1996; Molle et al. 2009). The project was justified by the Prime Minister in 1989 as a project "to turn the battlefields [of Indochina] into marketplaces" (Molle et al. 2009:260). This was conceived as a domestic project, where water from the Kong (Mekong) River was diverted into the Chi-Mun basin.

5.3.3 Hydraulic Mission and Reflexive Water Management

The Thai water policy toward the end of the IMC and during the MRC is at best expressed as a bricolage of the hydraulic mission and reflexive management. First, the hydraulic mission continued to be implemented in the northeast by the central government under the premise of "regional development." Feasibility studies of the KCM project were conducted by the National Energy Agency in 1992. The Rasi Salai weir and Huana weir were constructed and completed between 1992 and 2000, and projects on the tributaries of the lower Mun River were also completed (Floch et al. 2007). Despite the slow progress of the project, the hydraulic mission utilizing the mainstream waters is evident in the northeast region. Furthermore, the concept of the Water Grid was proposed by the then Prime Minister in 2003. The Ninth National Economic and Social Development Plan aimed to increase irrigation area by approximately 13% compared to the previous plan (Budhaka et al. 2002). To achieve these goals, the Water Grid planned several transbasin diversions and diversions from Cambodia and Laos, costing roughly USD 5 billion (200 billion baht). Irrigation area would be increased immediately to 17 million ha in just 5 years, and the northeast would gain the most under this project (Molle and Floch 2008). In 2004, the Royal Irrigation Department also proposed a capital-intensive project to increase irrigation area to approximately 21 million ha over 60 years using a network of pumps (Molle and Floch 2008). Despite these projects being highly promoted by key governmental officials and agencies, much of the plans remained on paper when the government changed in 2006 (Molle and Floch 2008).

At the same time, at least rhetorically, reflexive management was beginning to be emphasized in national water policies. For example, the Sixth National Economic and Social Development Plan (1987–1991) began to advise using water in a more economic way (Government of Thailand 1986). The next Seventh National Economic and Social Development Plan (1992–1996) encouraged the development of guidelines for water resource management in the direction of integrated management (Government of Thailand 1991). Some river basin organizations have been established as a result (Molle 2005). In the Eighth National Economic and Social Development Plan (1997–2001), integrated management of natural resources is further emphasized through its "Development of Popular Governance" (Government of Thailand 1996). Public participation in decision making is seen as an important feature of governance, thereby reflecting some of the principles of reflexive water management. In 2000, the National Water Vision was published with a distinct IWRM approach. The policy shift toward sustainable, holistic water management is also evident in the Ninth National Economic and Social Development Plan (2002–2006) (Government of Thailand 2001). Based on these policies, Thailand began to reform the water sector and decentralize water management (Molle 2005).

The emphasis on inclusive decision making and economizing water use has come about in a climate where there were increasing public protests and criticism of environmental projects. Gradually from the 1970s, civil society began voicing its concerns over environmental issues such as land tenure (Foran 2006), with the northeast becoming the center stage for some large-scale and successful protests of dam development in the 1980s and 1990s. The Nam Choam dam project in the northeast region is seen as a successful case of social activism effectively pressuring the central government to revise its plans (Rigg 1991). The Pak Mun dam project, also in the Chi-Mun basin, is also characterized by successful antidam movements (Foran 2006). The above-mentioned KCM project was by no means undisputed: local communities, activists, and academics have criticized the project (Molle and Floch 2008).

5.4 Impact of Domestic Water Policies on Multilateral Water Resources Development

The previous section examined the development of the hydraulic mission with a particular focus on the northeast region of Thailand. By doing so, the subsections touched upon the characteristics and development of Thai water policy. The political economy of water changed greatly during the half century after the Second World War, and water policy has emphasized different priorities in water resource management. The capital-intensive hydraulic mission attests to the power of the hydrocracy to mobilize funds and resources within the country on the premise of regional planning, rural development, and irrigation development. These water policies existed within a larger context of the Lower Mekong River basin development. In many cases, projects within the northeast were part of the international Mekong water management institutions' development plans. The following subsections analyze how the domestic water development and management policies relate to the politicization and securitization of international transboundary water resource development during the MC, IMC, and MRC periods.

5.4.1 Impact of Domestic Water Policies During the MC Period

When the MC was established, the premise of the multilateral platform was to develop the unharnessed Mekong waters. The concept of the MC was inspired by the Tennessee Valley Authority and thus focused on executing the hydraulic mission of the basin. The initiative by the UNECAFE to establish an institution for river development provided Thailand with an opportunity to pursue investment in large-scale hydraulic infrastructure for economic development. Because the MC activities were supported by international aid, there were many studies and projects that were achieved because of the mobilization of external expertise and capital – resources that postwar developing countries would have lacked.

Water resource development of the MC was based on managing water availability of the Lower Mekong River basin caused by the monsoon climate. By controlling water scarcity and flooding, economic development of the lower Mekong region would be increased. The hydraulic mission of the MC was convenient for Thailand who was also progressing with its hydraulic mission. The hydraulic mission in the northeast region was facilitated by the projects being devised under the MC. The Thai delegation to the MC enthusiastically supported water resource development by declaring:

> Part of the reply [to the question regarding the importance of the Mekong River to Thailand] is that Mekong electricity, and also increased agricultural production from the Mekong Scheme, will benefit the entire country, both directly -- electricity and food can come from the basin to Thai areas beyond the basin -- and indirectly, through aiding the entire economy including, we hope, our export potential. The other part of the answer concerns numbers of people. The North East of Thailand -- the area which stands to gain directly from the development of Mekong mainstream and tributary projects -- has as of now a population of 8 million people. Moreover they are the section of Thailand standing in greatest need of, and having the greatest desire for, economic improvement. Now 8 million people are quite a few, as far as Mekong riparian countries are concerned. And so, from the point of view of persons living within the Mekong basin, Thailand is quite as fully concerned as any other lower Mekong riparian. Hence our interest in the Mekong Project (MC 1962: Annex IX 44–45).

This speech by Boonrod Binson, a Thai representative to the MC, underlined the need to alleviate poverty in northeast Thailand and the basin on the whole.

In particular, mainstream development was considered crucial for economic development. In the same speech, Binson described the Pa Mong dam as a "basic requirement" for the development of the region (MC 1962: Annex IX 45). A report described the Pa Mong dam as the following:

> One single dam, such as Pa Mong could provide a massive block of power to meet essential needs so cheaply that net savings of some 100, 000, 000 dollars per year could be realized. This is a staggering sum, but those who think of the fate of the ever-increasing population of the northeast [region of Thailand] would probably find it even more important that in the future only Mekong water, stored behind a mainstream dam, could sustain the irrigation development needed for that region to produce enough food, rather than becoming a liability to the rest of the country (MC 1977:95).

The Pa Mong dam is described as a project that is vital to the survival of the local communities of the northeast ("fate of the ever-increasing population of the northeast") and the solution to the problem of economic underdevelopment. During the MC period, Thailand "astutely linked manipulation of Mekong water for the benefit of Thai development aspirations to the continuing participation ... in the evolving cooperative forum of the Mekong Committee" (Sneddon 2003:2240). The hydrocracy of Thailand successfully politicized water development in the northeast region to further facilitate investment in hydraulics through the MC.

5.4.2 Impact of Domestic Water Policies During the IMC Period

The Thai hydraulic mission was deeply related to the wider Mekong water resource development of the MC, but in the interim state of the committee, the hydrocracy could not rely on mainstream projects. As a result, Green Isaan and the KCM project were proposed as domestic projects under domestic water policies. Floch et al. (2007:27) commented that though the two-pronged water policy incorporates small-scale measures for securing water supply, the policy "merely reshuffled and reordered preferences towards completion of existing infrastructure and implementation of small-scale developments in the short-run, while retaining the long-term vision of the needs for irrigation development." Understanding this long-term vision is the key to analyzing subsequent actions of Thailand in the multilateral forum. While the two large-scale projects were domestic in nature, they utilized the mainstream water through diversions and transfers. Since using mainstream waters affects downstream flow, it could be argued that the projects need to be consulted with other basin states under the *Joint Declaration of the Principles for Utilization of the Waters of the Mekong Basin* established in 1975 by the MC. However, this declaration was never legalized, and the IMC did not have any clear guidelines for water use. The *Declaration Concerning the Interim Committee for Coordination of Investigations of the Lower Mekong Basin*, signed in 1978 when the IMC was established, lacked specific legal rules on water allocation and utilization. The ambiguity of water use gave rise to a situation where there was no clear distinction between national projects in the basin and multilateral projects belonging to the IMC (IMC: 1983b). The large-scale domestic waters proposed during this period were based on a long-term vision for water security that was unaffected by specific water allocation guidelines or rules of transboundary river basin management.

Thus, it is no surprise that when Cambodia requested to rejoin the committee in 1991, the issue of rules regarding water allocation and development was sensitive and consumed much of the negotiations of the institutional setup. Initially, it was intended that original documents of the MC (Statute of 1957 and the Joint Declaration of 1975) would be used when the commission resumed as a quadrilateral organization. However, there were serious differences of view on this matter between Thailand and Vietnam. Vietnam argued that the original rules of the MC should be revived. Thailand insisted that the rules be replaced because they had become outdated (Makim 2002) and maintained that it had the right to extract mainstream water equal to the quantity its tributaries contributed (Weatherbee 1997). The 1957 Statute and 1975 Joint Declaration required unanimity on development project both on the mainstream and tributaries. This effective "veto right" was seen as an impediment for future Thai water resource development plans (Nakayama 1999). With projects like the KCM on the national agenda, Thailand had the incentive to safeguard its plans for utilizing the mainstream flow and limit intervention on projects by other basin states.

Thailand's "uncompromising" position (Weatherbee 1997:175) is further illustrated in the securitizing move of water allocation and utilization. While negotiations were being conducted between the four states, the Thai Foreign Minister unilaterally proposed and hosted a meeting with all basin states, including China and Myanmar, and without the Commission secretariat. This action was out of bounds of the usual MC politics since the upstream countries had always been excluded, and the Commission secretariat was a crucial player in the multilateral negotiations. By taking an extraordinary measure (Buzan et al. 1998), the Thai hydrocracy closed off political space in which negotiations over water allocation rules would be done within the usual MC framework, thus securitizing the issue. Vietnam did not participate in this meeting in protest, and the relationship between the Thai governmental officials and the executive agent of the MC deteriorated (see Browder 1998). This incident shows that Thailand's national water development plans, which carried much political prestige, had an influence on the way the Thai hydrocracy formed its "national" position during the international negotiations over water allocation rules.

Even though the situation was diffused with the United Nations Development Programme (UNDP) intervening to facilitate the negotiations, the issue of water allocation and utilization was still heavily politicized. The negotiations were a political platform in which the Thai hydrocracy proposed and counterproposed texts of clauses to determine water utilization on the mainstream and tributaries (see Radosevich 1995). The Thai position was to require minimum disclosure and multilateral consent on projects using both the mainstream and tributary waters. For example, Thailand maintained that prior consultation would suffice for dry-season mainstream use, instead of more restrictive measures such as prior agreement (Radosevich 1995).

In the end, when the *Agreement on the Cooperation for the Sustainable Development of the Mekong River Basin* was signed in 1995, the water use principles were less restrictive compared to the 1975 Joint Declaration. The article concerning "Reasonable and Equitable Use" (Mekong Agreement 1995 Article 5) conceded to Thailand's concerns about limiting control over water development projects (Nakayama 1999; Makim 2002). In this article, tributary uses are subject to prior notification. Wet-season mainstream uses are subject to notification and prior consultation for intrabasin use and interbasin diversion, respectively. For dry-season mainstream uses, where quantitative allocation is particularly sensitive, prior consultation "which aims at arriving at an agreement" (Article 5) is required for intrabasin use and interbasin diversion, respectively. Under these articles, the KCM project would be considered a tributary project that requires notification only, thereby allowing Thailand to have the option of developing the project with comparatively less restriction than rules agreed in previous water management regimes.

The progress of water resource development in the form of domestic projects utilizing the Mekong waters heavily influenced the Thai hydrocracy's position during the negotiations for a new multilateral institution. New rules had to be agreed over water allocation and utilization. The hydrocracy securitized and politicized this issue of water utilization with an aim to formulate rules and procedures that would safeguard its domestic projects.

5.4.3   Impact of Domestic Water Policies During the MRC Period

Water policy reform in Thailand has, at least on paper, focused on the integrated and decentralized principles of IWRM. Policy-wise, this explicit focus on IWRM does not contradict the regional efforts of IWRM facilitated by the MRC during its initial years until 2006. The Thai delegation officially expressed interest and support for the IWRM approach in the MRC, citing its potential for poverty alleviation (see Statement by Suwit Khunkitti, in MRC 2004). In order to achieve sustainable development, the 1995 Agreement stipulates that Basin Development Plans (BDP) are the "tools and process the Joint Committee would use as a blueprint to identify, categorize and prioritize the projects and programs" (Mekong Agreement 1995). The first phase of the BDP (2000–2006) was distinctly characterized by IWRM principles and produced a document, "Strategic directions for IWRM in the Lower Mekong Basin." The assumption of adopting IWRM is that water utilization will be made efficient with the inclusion of a wider stakeholder base and will assist poverty alleviation (MRC 2006). The first phase has only laid the foundation for implementing IWRM, and effects of the Plan will take time to realize the "sustainable" use of the Mekong waters. However, this idea of inclusive decision making has been taken up by the Thai hydrocracy that has borne an interesting situation in the dynamics of the basin states. As one form of integrated decision making, the Thai hydrocracy has argued for the involvement of upstream China and Myanmar in the MRC. China and Myanmar are not part of the MRC but participate as dialogue partners. Thailand has been keen to involve these upstream countries based on the argument that for the maintenance of mainstream flow, basin-wide planning of both the upstream and downstream basins is needed and that the involvement of China and Myanmar would be vital for future water management (MRC 2005). By calling for the inclusion of China and Myanmar, Thailand is opening up new political space in the multilateral fora to question the basic rules that bind the MRC regime, thereby politicizing water allocation and utilization on the mainstream, once again.

5.5   Discussion and Conclusion

This chapter presented a historical analysis of the impact of domestic water policies on international transboundary water development and management. While the chapter does not make any predictions about future challenges or successes in the practice of transboundary IWRM in the Mekong River basin (after all, transboundary IWRM is still relatively new to the region), there are some important policy implications that can be gleaned from looking at the Thai case study.

First, the way Thai water policy has changed over time shows that the hydraulic mission does not give away easily to reflexive modes of management. The discrepancy between water policy principles and implementation is not unique to Thailand.

For example, Swatuk (2008) showed that in actuality, different paradigms exist in parallel in Southern Africa. Swatuk argued that there are two main camps operating in the water resources development landscape: the technocentric coalition of those with vested interests in the hydraulic mission and the ecocentric coalition of those promoting affordable water through environmentally friendly and participatory processes. In this landscape of divided stakeholders, the technocentric approach is more influential because "the hard path to water development is generally believed to facilitate economic development and so deliver jobs, votes, money, influence and power" (Swatuk 2008:41). The prominence of the technocentric coalition is also evident in Thailand, where the Water Grid is seen as the "ultimate avatar" of mega-project plans to bring water to underdeveloped areas of the northeast (Molle and Floch 2008:203).

Second, in a situation where there is a gap between IWRM principles and practice, it is likely that the overall effect of transboundary IWRM implemented by the MRC will not be as effective. As noted in the lessons from the first phase of the BDP, the level of IWRM implementation in the four riparian countries differs, and the involvement and contribution of line agencies in the countries were important (MRC 2006). It has been long pointed out that the policy implementing agencies in Thailand are numerous and that the coordination among them fragmented (Christensen and Boon-Long 1994). In Thailand, the hydraulic mission still captures the minds and resources of the hydrocracy despite the gradual introduction of IWRM measures. Some have been critical about the extent to which public participation will be achieved (see Molle 2005). Bandaragoda (2006) argued that the hydrocracy has strong administrative and legal powers, thus making the implementation of IWRM slow; this argument seems to be applicable for the case of Thailand.

Thirdly and related to the second point, efforts of transboundary IWRM by the MRC may be rendered ineffective if the basin states do not find the multilateral institution convenient for their needs. Hensengerth (2008) analyzed that the MRC is used by the riparian countries to gain side benefits and not necessarily to produce regional public goods of basin-wide water management. As a result, other fora are becoming attractive to the basin states, including Thailand to pursue hydraulic development in the Mekong, while still engaging in the MRC. For example, while not yet having an individual water sector component (but facilitating navigation and energy issues), the Greater Mekong Subregion proposed by the Asian Development Bank has initiated infrastructure development in a wider geographical scope, which includes upstream China. The recent hydropower developments on the mainstream also provide Thailand with an incentive to invest in hydraulic missions of other states such as Laos on a bilateral basis, without the involvement of the MRC. This situation not only further questions the governance capacity of the MRC and its role in water management (see Hirsch et al. 2006; Affeltranger 2009) but also shows that the MRC is not as attractive to Thailand as the MC, when major hydraulic investments had yet to be made.

The analysis above has shown instances where Thai delegations to the multilateral river basin committees have politicized and securitized the issue of water allocation and utilization. Consequently, it was shown that Thai domestic water policies have informed the negotiating positions of the Thai hydrocracy and the "national" interests as represented by them in the three multilateral water management institutions, thereby influencing transboundary water management and development. The analysis only focused on the Thai hydrocracy as an actor that develops and utilizes domestic water policy. In order to deepen the analysis, the role of other actors such as the civil society in Thailand can be examined in understanding their influence on water resource development progresses and changes at the national level, and their influences on international transboundary river management and development. Furthermore, this study can be complemented by understanding the factors other than domestic water policies, such as the influence of development by the upstream states and of the donor states and organizations to the Mekong basin region that contribute to transboundary river management and development.

References

Affeltranger B (2009) Sustainability of environmental regimes: the Mekong river commission. In: Brauch HG, Spring ÚO, Grin J, Mesjasz C, Kameri-Mbote P, Behera NC, Chourou B, Krummenacher H (eds) Facing global environmental change: environmental, human, energy, food, health and water security concepts. Springer, Berlin/Heidelberg, pp 593–601

Allan JA (2003) IWRM/IWRAM: a new sanctioned discourse? SOAS/KCL water issues group occasional paper 50. SOAS/King's College London, London

Bandaragoda DJ (2006) Status of institutional reforms for integrated water resources management in Asia: indications from policy reviews in five countries. Working paper 108. International Water Management Institute (IWMI), Colombo, Sri Lanka

Behrman JR (1968) Significance of intracountry variations for Asian agricultural prospects: central and northeastern Thailand. Asian Surv 8(3):157–173

Browder G (1998) Negotiating an international regime for water allocation in the Mekong river basin. PhD thesis, Stanford University

Browder G, Ortolano L (2000) The evolution of an international water resources management regime in the Mekong river basin. Nat Resour J 40(3):499–531

Budhaka B, Srikajorn M, Boonkird, V (2002) Thailand country report on investment in water. In: Investment in land and water. Proceedings of the regional consultation, Food and Agriculture Organization of the United Nations, Bangkok, pp 325–337

Buzan B, Wæver O, de Wilde J (1998) Security: a new framework for analysis. Lynne Rienner Pub, Boulder

Chomchai P (1994) The United States, the Mekong committee and Thailand: a study of American multilateral and bilateral assistance to north-east Thailand since the 1950s. Institute of Asian Studies, Chulalongkorn University, Bangkok

Christensen SR, Boon-Long A (1994) Institutional problems in Thai water management working paper for natural resource and environment program, Thailand Development Research Institute Foundation, Bangkok

Declaration concerning the interim committee for coordination of investigations of the lower Mekong basin, signed by the representatives of the governments of Laos, Thailand and Vietnam to the committee for coordination of investigations of the lower Mekong basin at Vientiane, 5 Jan 1978

Dixon CJ (1999) The Thai economy: uneven development and internationalisation. Routledge, London/New York

Donner W (1978) The five faces of Thailand: an economic geography. C. Hurst/Institute of Asian Affairs, Hamburg, London

Feeny D (1982) The political economy of productivity: Thai agricultural development, 1880–1975. University of British Columbia Press, Vancouver

Floch P, Molle F, Loiskandl W (2007) Marshalling water resources: a chronology of irrigation development in the Chi-Mun river basin, northeast Thailand no. M-POWER working paper MP-2007-02. Chiag Mai University, Unit of Social and Environmental Research

Foran T (2006) Rivers of contention: Pak Mun dam, electricity planning, and state–society relations in Thailand, 1932–2004. PhD thesis, University of Sydney

Government of Thailand (1986) Sixth national economic and social development plan: 1987–1997. Government of Thailand, Bangkok

Government of Thailand (1991) Seventh national economic and social development plan: 1992–1996. Government of Thailand, Bangkok

Government of Thailand (1996) Eighth national economic and social development plan: 1997–2001. Government of Thailand, Bangkok

Government of Thailand (2001) Ninth national economic and social development plan: 2002–2006. Government of Thailand, Bangkok

GWP (Global Water Partnership) (2000) Integrated water resources management no. TAC background papers no 4. Global Water Partnership, Stockholm, Sweden

Hensengerth O (2008) Transboundary river cooperation and the regional public good: the case of the Mekong river. Paper presented at the 13th IWRA world water congress, International Water Resources Association, Montpellier, 1–4 Sept 2008

Hirsch P, Jensen KM, Boer B, Carrard N, FitzGerald S, Lyster R (2006) National interests and transboundary water governance in the Mekong. Australian Mekong Resource Centre/In collaboration with Danish International Development Assistance and the University of Sydney, Sydney

Hooper BP (2003) Integrated water resources management and river basin governance. Water Resour Update 126:12–20

Hori H (1996) Mekong-gawa: Kaihatsu to kankyou. Kokon-Shoin, Tokyo

IMC (Interim Committee for the Coordination of Investigations of the Lower Mekong Basin) (1983a) Mainstream development: supplementary Pa Mong studies. Note by the executive agent. IMC, Bangkok

IMC (Interim Committee for the Coordination of Investigations of the Lower Mekong Basin) (1983b) Mainstream development and related issues: note by the executive agent. IMC, Bangkok

IMC (Interim Committee for the Coordination of Investigations of the Lower Mekong Basin) (1985) Thailand and the Mekong project: note by the executive gent. IMC, Bangkok

Jacobs JW (1995) Mekong committee history and lessons for river basin development. Geogr J 161(2):135–148

Joint declaration of principles for utilization of the waters of the lower Mekong basin, signed by the representatives of the government of Cambodia, Laos, Thailand and Vietnam to the committee for coordination of investigations of the lower Mekong basin at Vientiane, 31 Jan 1975

Kaida Y (1978) Irrigation and drainage: present and future. In: Ishi Y (ed) Thailand: a rice growing society (Revised translation of Tai-koku, published in 1975 ed), University Press of Hawaii, Honolulu, pp 205–245

Keskinen M, Mehtonen K, Varis O (2008) Transboundary cooperation vs. internal ambitions: the role of China and Cambodia in the Mekong region. In: Pachova NI, Nakayama M, Jansky L (eds) International water security: domestic threats and opportunities. UNU Press, Tokyo

Makim A (2002) Resources for security and stability? the politics of regional cooperation on the Mekong, 1957–2001. J Environ Dev 11(5):5–52

MC (Committee for Co-ordination of Investigations of Lower Mekong Basin) (1962) Seventeenth Session (Special) 7–26 Mar 1962 Tokyo, Japan. MC, Bangkok

MC (Committee for Co-ordination of Investigations of Lower Mekong Basin) (1977) Thailand and the Mekong project: information note by the secretariat. MC, Bangkok

Mekong Agreement (1995) Agreement on the cooperation for the sustainable development of the Mekong River Basin at Chiang Rai, Thailand, 5 Apr 1995

Molle F (2005) Irrigation and water policies in the Mekong region: current discourses and practices research report no 95, IWMI, Colombo

Molle F, Floch P (2008) Megaprojects and social and environmental changes: the case of the Thai "water grid". Ambio 37(3):199–204

Molle F, Floch P, Promphakping B, Blake DJH (2009) The 'Greening of Isaan': politics, ideology and irrigation development in the northeast of Thailand. In: Molle F, Foran T, Käkönen M (eds) Contested waterscapes in the Mekong region: hydropower, livelihoods and governance. Earthscan, London/Sterling V.A, pp 253–282

Mollinga PP (2008) Water, politics and development: framing a political sociology of water resource management. Water Altern 1(1):7–23

MRC (Mekong River Commission) (2004) Minutes of the eleventh meeting of the council, Mekong River Commission, Vientiane, Lao PDR, 8–9 Dec 2004

MRC (Mekong River Commission) (2005) Minutes of the twelfth meeting of the council. 30 Nov–1 Dec 2005, Chiang Rai, Thailand. Mekong River Commission, Vientiane

MRC (Mekong River Commission) (2006) Basin development plan completion report for phase 1 2001–2006. Mekong River Commission, Vientiane

Nakayama M (1999) Aspects behind differences in two agreements adopted by Riparian countries of the lower Mekong river basin. J Comp Policy Anal Res Pract 1(3):293–308

Öjendal J (2000) Sharing the good: modes of managing water resources in the lower Mekong river basin. Department of Peace and Development Research, Göteborg University, Gothenburg

Radosevich G (1995) Agreement on the cooperation for the sustainable development of the Mekong river basin: commentary and history. UNDP, Bangkok

Rigg J (1991) Thailand's Nam choan dam project: a case study in the 'Greening' of south- east Asia. Glob Ecol Biogeogr Lett 1(2):42–54

Rittel HWJ, Webber MM (1973) Dilemmas in a general theory of planning. Policy Sci 4(2):155–169

Roe P (2006) Reconstructing identities or managing minorities? desecuritizing minority rights: a response to jutila. Secur Dialogue 37(3):425–438

Statute of the committee for co-ordination of investigations of lower Mekong basin established by the governments of Cambodia, Laos, Thailand and the other republics of Viet-Nam in response to the decision taken by the united nations economic commission for Asia and the Far East, Phnom-Penh (Cambodia), 31 Oct 1957

Sneddon C (2003) Reconfiguring scale and power: the Khong-Chi-Mun project in northeast Thailand. Environ Plann A 35(12):2229–2250

Swatuk LA (2008) A political economy of water in southern Africa. Water Altern 1(1):24–47

Tingsanchali T, Singh PR (1996) Optimum water resources allocation for Mekong-Chi-Mun transbasin irrigation project, northeast Thailand. Water Int 21(1):20

Warner J (2000) Global environmental security: an emerging 'concept of control'. In: Stott P, Sullivan S (eds) Political ecology: science, myth and power. Arnold, London, pp 247–265

Weatherbee DE (1997) Cooperation and conflict in the Mekong river basin. Stud Confl Terror 20(2):167–184

Wester P (2008) Shedding the waters: institutional change and water control in the Lerma-Chapala basin, Mexico. PhD thesis, Wageningen University

United Nations Survey Mission (1958) Program of studies and investigations for comprehensive development lower Mekong river basin. United Nations, Bangkok

Chapter 6
Mekong at the Crossroads: Alternative Paths of Water Development and Impact Assessment

Marko Keskinen, Matti Kummu, Mira Käkönen, and Olli Varis

Abstract The Mekong Region in Southeast Asia is undergoing rapid transitions socially, economically, and environmentally. Water is related to these changes in a very profound manner, and the Mekong River and its tributaries are seeing increasing number of plans for water development, most notably in the form of large-scale hydropower. The impacts of this development vary among regional, national, and local levels and across different timescales, influencing societies, politics, and the environment in a variety of ways. While different impact assessment and water management frameworks – including Integrated Water Resources Management (IWRM) – have been used by actors at different levels in the basin, they have not been too successful in analyzing and communicating the various development paths and their differing impacts in all their complexity. This chapter discusses the water development pathways in the Mekong Basin, including their potential impacts and the different possibilities to assess them, as of early 2010. It is concluded that the water development and related management practices in the Mekong are at the crossroads methodologically and, even more importantly, politically.

M. Keskinen (✉) • M. Kummu • O. Varis
Water & Development Research Group, Aalto University, P.O. Box 15200, FIN-00076
Aalto, Finland
e-mail: marko.keskinen@aalto.fi; matti.kummu@aalto.fi; olli.varis@aalto.fi

M. Käkönen
Water & Development Research Group , Aalto University , P.O. Box 15200, FIN-00076
Aalto , Finland
Finland Futures Research Centre, University of Turku, Korkeavuorenkatu 25 A 2, FIN-00130
Helsinki, Finland
e-mail: mira.kakonen@utu.fi

J. Öjendal et al. (eds.), *Politics and Development in a Transboundary Watershed:*
The Case of the Lower Mekong Basin, DOI 10.1007/978-94-007-0476-3_6,
© Springer Science+Business Media B.V. 2012

6.1 Introduction: The Changing Mekong

The Mekong Region is undergoing rapid transitions socially, economically, and environmentally. Economies of the Mekong countries are stabilizing after the political turbulence of several decades, and development pressures towards the region's natural resources are vast. Water is related to these changes in a very profound manner, and the Mekong River and its tributaries are seeing an increasing number of plans for water development. The most remarkable, and most fervently debated, element of such plans is the development of large-scale hydropower. They are therefore also the main focus of this chapter. However, also several other changes, including intensification of agriculture, construction of new infrastructure, and changes in the land use, are likely to have notable impacts to the Mekong's waters. The impacts of these developments vary among regional, national, and local levels and across different timescales, influencing societies, politics, and the environment in a variety of ways. At the same time, new driving forces, most importantly climate change, are entering the discussion, affecting the ways the water resources are being used and developed. Decisions about the forms of water development will therefore have profound and far-reaching implications – not only physical and ecological but also social and political – throughout the basin.[1]

While various impact assessment and water management frameworks, such as Integrated Water Resources Management (IWRM), have been used and adapted by the actors at different levels in the Mekong River Basin, they have not (yet) been too successful in analyzing and communicating the various development paths and their differing impacts in all their complexity. Indeed, while the drive for increased utilization of the river's waters is intensive in all riparian countries, the understanding of the actual impacts of these developments is in many aspects vague, and discussion about the most sustainable development options remains weak. The situation is, however, improving, and there exist an increasing number of initiatives that study and discuss the potential impacts – and general feasibility – of current water development plans. Such initiatives range from water dialogues carried out by actors, such as the M-POWER network (IUCN et al. 2007a), to the "IWRM-Based Basin Development Strategy" and related assessments implemented by the regional Mekong River Commission (MRC 2009a, b).

This chapter discusses the water resources development pathways in the Mekong Basin and considers the different possibilities to assess their impacts. By presenting examples of potential impacts on water quantity, quality, and ecosystem productivity, we seek to highlight the diversity of impacts that water development is likely to induce at different scales. Such examples illustrate the first dimension of the crossroads related to possible development paths and their impacts. The chapter also discusses the challenges related to current practices of impact assessment and water resources management, addressing therefore the second, methodological dimension of the crossroads: the choice between different approaches used for management and impact assessment.

---

[1] The first full version of this chapter was submitted for review in February 2009 and the updated version in February 2010. Consequently, some of the discussion presented may be partly out-of-date due to rapid progress of Mekong's hydropower plans and related assessments.

The diverse set of impacts presented in this chapter emphasizes the fundamental threats to water resources in the basin, underlining the often neglected importance of fisheries, floodplains, and other common pool resources. Based on this, we argue for management and planning processes that build on existing livelihoods and resource uses, rather than on projects seeking to replace them (see, e.g., Öjendal 2000; Hirsch et al. 2006; MRCS/WUP-FIN 2007; Varis et al. 2008). We emphasize the need for parallel processes and methods for management and impact assessments. Indeed, due to complexities and uncertainties involved, there is a need to use a set of different models and impact assessment methods when assessing the diverse impacts to the water-related ecosystems and livelihoods. This can also be seen to pose a challenge to the concept of IWRM that – despite its calls for context-specificity – tends often to highlight a common, relatively predefined management approach for different scales and contexts.

Finally, while drawing on the analogy of crossroads, we acknowledge the limitations of such a view. In reality, the crossroads does not necessarily exist, at least not as one simple concrete and desirable choice to be made at a particular point in time. For the decisions about the water development – and the methods used in related planning and assessment – are usually done continually through a political process involving actors with varying agendas and interests. Consequently, the decision making about certain types of development has often at least as much to do with power structures and specific ideals than with balanced assessment of different alternatives. We do hope, however, that by highlighting the alternative paths that such decisions could take, we contribute for the broader discussion about the possible ways to use and develop the Mekong's waters.

6.2 Water Development in the Mekong

The human impact on water resources has increased dramatically during the last decades all over the world (Vörösmarty and Sahagian 2000). The Mekong River is one of the few large river basins in the world that has not been irreversibly modified by large-scale infrastructure. While the first dams in the Mekong mainstream (upstream in China; see Chap. 9 of this volume) and several dams in the tributaries have already been built, flow regimes in the lower reaches of the mainstream are still, essentially, natural (MRC 2005).[2] These conditions may not last much longer, as the Mekong River Basin is facing the prospects of a major boom in water infrastructure projects. Huge hydropower dams as well as water diversions for irrigation are planned in different parts of the basin, some on tributaries and others on the mainstream (King et al. 2007; MRC 2008a, 2010). As hydropower dams are expected to have the most radical impacts for the river flows and related ecosystems, they are next discussed in more detail.

There are currently various plans for hydropower development in the Mekong Basin. It is thereby challenging to just keep track of all of them, and even more

---

[2] Strongly modified waterscapes are also found within the basin: the Mekong Delta of Vietnam is a particularly interesting example of water regime intensively regulated by human interventions (see, e.g., Biggs 2004; Miller 2006; Käkönen 2008).

Table 6.1 Estimates for existing, ongoing, and proposed hydropower dams projects in six Mekong countries (Modified from King et al. 2007)

	Existing dams		Under construction		Proposed/ potential	
	Total	Mekong	Total	Mekong	Total	Mekong
Cambodia	3	1	1	0	33	26
Yunnan, China	2	2	3	2	34	10
Laos	8	8	3	3	32	32
Myanmar	13	0	8	0	15	1
Thailand	10	10	1	0	0	0
Vietnam	18	9	12	9	65	9
Total	54	30	28	14	179	78

Figures for all dams plus those within the Mekong Basin

challenging to estimate the cumulative impacts that such plans are likely to have. Two prominent sources, however, provide some general estimates about the scale of current hydropower plans (King et al. 2007; MRC 2008a). A recent inventory of existing and potential hydropower projects in the six Mekong countries came up with a total of 261 hydropower projects in the region, including 122 projects within the Mekong River Basin (King et al. 2007). Out of this total, an estimated 14 projects were under construction and a further 78 large projects were identified as potential sites within the basin (Table 6.1).

In autumn 2008, the Mekong River Commission (MRC) published a map indicating the location of dams planned in the Lower Mekong Basin (Fig. 6.1). When combined with available information from China, this data includes 28 existing hydropower dams as well as an estimated 14 dams that are under construction, and additional 101 dams that are at the planning stage, most of them in Laos (MRC 2008a). Notable is that this MRC data indicates plans for mainstream dams also outside China, including eight dams in Laos, two in Cambodia, and one in the border area between Laos and Thailand (MRC 2008b). These would be the first dams to be located in the Lower Mekong mainstream and also first mainstream dams to be constructed by an MRC member country. Such plans have thus put also the Mekong River Commission into a new position. Indeed, to respond to these new plans, the MRC has already strengthened its impact assessment practices, including Strategic Environmental Assessment (SEA) of the proposed mainstream dams as well as detailed assessments looking at hydrological, environmental, social, economic, and fish-related impacts of various different water development scenarios under MRC's Basin Development Plan (MRC 2009b, c, 2010).

Total theoretical potential for hydropower production in the entire Mekong Basin has been estimated to be around 43,000 MW, with some 30,000 MW technically available[3] in the four Lower Mekong Basin countries either in the

---

[3] According to WEC (2007), technically exploitable hydropower capability (namely potential) is the amount of the gross theoretical capability that can be exploited within the limits of current technology. Economically exploitable capability, on the other hand, is the amount of the gross theoretical capability that can be exploited within the limits of current technology under present and expected local economic conditions.

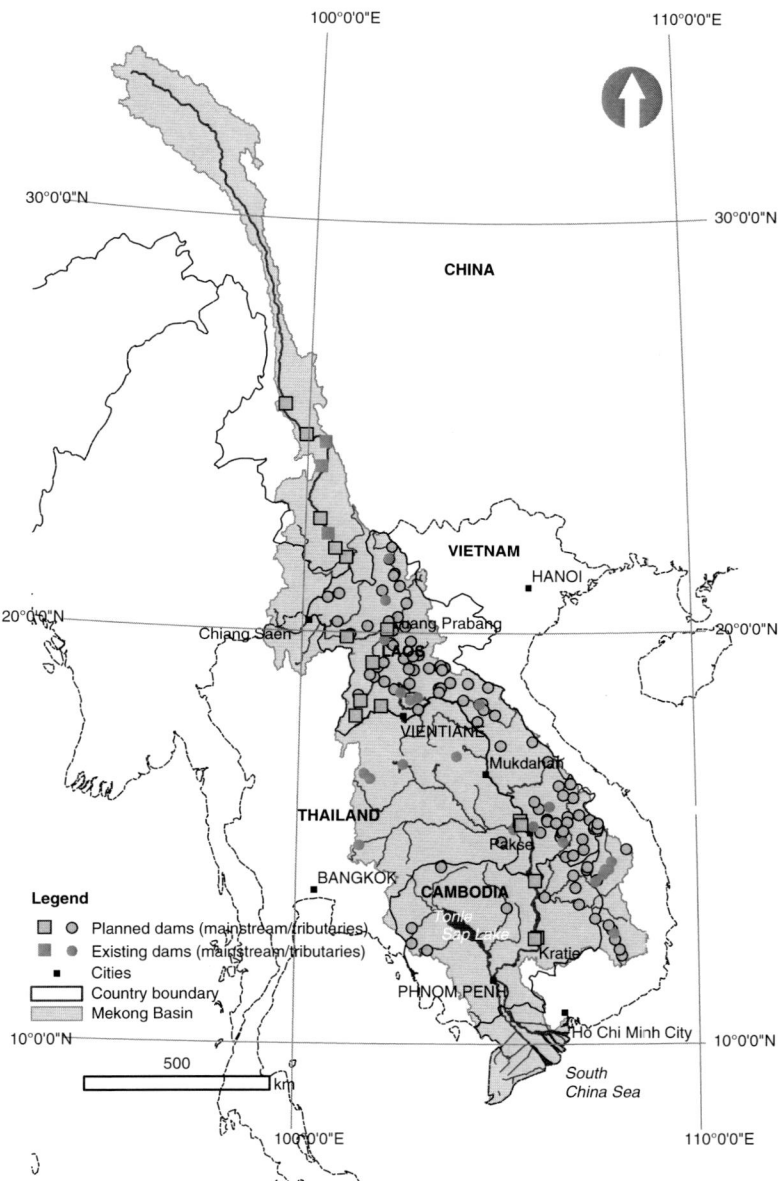

Fig. 6.1 The map on existing and planned hydropower dams in the Mekong Basin, showing existing (*darker*) and planned (*lighter*) projects (Modified from MRC 2008a)

mainstream (13,000 MW) or tributaries (17,000 MW) (King et al. 2007). Considerable amount of both the production capacity and active storage capacity is located in the Chinese part of Lancang-Mekong River, with plans including construction of projected 15,600 MW in the Mekong mainstream with a combined active storage of 23,200 million cubic meters by year 2025 (King et al. 2007). It is therefore relatively clear that next decade is likely to see an increasing amount of large-scale hydropower development both in the upper and lower parts of the basin. The actual cumulative impacts of these dams will depend on their amount, location, and storage capacity, as well as operational procedures, but the impacts are in any case most likely to be major. Consequently, looking at the current development plans and the overall political economy of the basin, we are likely to see a paradigmatic change in the water development in the entire region. Such developments – and consequent impacts – will also place new expectations for the water management practices at both national and regional level.

6.3   Estimating the Impacts

One of the central aspects of any decision-making process related to water development is to estimate the potential impacts that such development is likely to have. Consequently, different impact assessment methods are being increasingly used to inform water development planning. Indeed, decision making is relying nowadays so much on technical expertise and assessments that Rayner (2003: 163) has characterized the present era as the "age of assessment."

The impacts of water development can be both positive and negative, and they can also be felt very differently in different areas and times as well as by different social groups. In terms of hydropower development, more secure electricity production, increased water availability and predictability for agriculture, and income gained from the export of hydroelectricity represent obvious and often-stated examples of positive impacts, and ultimate reasons, for such projects. At the same time, however, there are also a variety of environmental, social, and economic impacts that the water development causes to water-related ecosystems, and consequently on livelihoods and industries dependent on them.

Numerous impact assessment processes have been undertaken also in the Mekong River Basin by actors at various levels. A great majority of these assessments indicate that the planned water developments in the basin are likely to cause remarkable changes for the availability of water-related resources – most importantly fish – and, consequently, for the livelihoods and food security of millions of people (see, e.g., MRC 2006a, 2009c, 2010; IUCN and IWMI 2007a; MRCS/WUP-FIN 2007; Dugan 2008). Yet, the estimates about the actual magnitude of such impacts remains varied, with different assessments providing widely differing estimates on the potential environmental, social, and economic impacts. Particularly basin-wide assessments have several challenges related to their comprehensiveness and overall reliability (see, e.g., Mirumachi and Nakayama 2007; MRCS/WUP-FIN 2007; Wyatt and Baird 2007; Keskinen 2008, 2010; Kummu and Sarkkula 2008).

6.4 Hydrological Models as a Tool for Impact Assessment

Various kinds of computational models provide one way to simulate the potential changes in the river system due to different kinds of developments.[4] Models are generally used to improve understanding of cumulative and aggregate effects, to provide forecasts, and to help to quantify different scenarios. These in turn are helpful for long-term planning of water resources development as well as for the assessment of water-related impacts. There has, however, also been active discussion about the challenges linked with the models and their results, related for instance to their transparency, reliability, and the possibilities for misuse (see e.g., Sarkkula et al. 2007; Käkönen and Hirsch 2009).

This chapter draws on the findings from the hydrological modeling and impact assessment work carried out in the Lower Mekong Modelling Project (WUP-FIN) under the Mekong River Commission (MRCS/WUP-FIN 2007). The hydrological models of the WUP-FIN Project used the basin-wide scenarios developed within the Decision Support Framework (DSF) of the MRC as their starting point to simulate the changes in flow regime with foreseen hydropower developments in subbasin scale. In addition, environmental and socioeconomic analyses were carried out to understand better the consequent environmental, social, and economic impacts that such flow changes are likely to have (see MRCS/WUP-FIN 2007; Sarkkula et al. 2007). However, as the scenarios used in the impact assessment of the WUP-FIN Project were developed already several years ago, even the most radical scenario, i.e., so-called High Development Scenario,[5] included only Chinese mainstream dams and some Lower Mekong Basin tributaries dams. Consequently, the estimates presented in this chapter can be considered to be relatively moderate, and the actual cumulative impacts of the current hydropower development plans – including several dams for the Lower Mekong mainstream – are likely to be much bigger in terms of changes in both water quantity and water quality.

---

[4] For more information on models and their use in environmental planning and decision making in the Mekong, see, e.g., Jakeman et al. (2006) and Sarkkula et al. (2007).

[5] The MRC used up to 2005 five development scenarios to assess the potential impacts of different development paths: Chinese Dams, Low Development, Embankments, Agriculture, and High Development (World Bank 2004). While such an approach was very useful in highlighting the differences between the estimates for different scenarios, the MRC later on gave up using them and replaced them with less controversial – and less illustrative – Flow Regimes (see also Käkönen and Hirsch 2009). More recently, however, the different development scenarios have been brought back to the discussion, and current MRC publications include several development scenarios for the Mekong, even with different timescales (MRC 2009a, b, 2010).

6.5   An Array of Impacts, Radical Consequences

This section discusses the projected impacts of large-scale hydropower development in the Mekong Basin by presenting examples of the impact estimates on water quantity, water quality, and ecosystem productivity. By doing this, we aim to highlight two important issues. First of all, the existing estimates already in some, relatively simple water-related indicators such as water levels and sediments point toward remarkable potential changes due to hydropower development. Second, the examples illustrate that the actual impacts to systems as complex as floodplains or fisheries are much more difficult to estimate, since the impacts to these systems are felt through a combination of several impacts, both direct and indirect. In addition, due to the critical social and economic importance of the floodplains and the fisheries, the physical and ecological impacts need to be closely connected with broader social and political dimensions – a process that is still at a very early stage in the Mekong. Consequently, we hope that the findings presented in this chapter are useful also when studying and discussing the ongoing impact assessment processes and their results, including those within the MRC (2009b, c, 2010).

Most of the discussion on the potential impacts presented in this chapter focuses on the Tonle Sap Lake system that forms a particularly important economic, social, and environmental resource for the entire Mekong Basin and for Cambodia in particular (Fig. 6.2). Overall, the Tonle Sap Lake and the resources it supports form a central source of livelihoods and food for well over a million people living in the lake and its floodplains (Keskinen 2006; Keskinen et al. 2007). The significance of

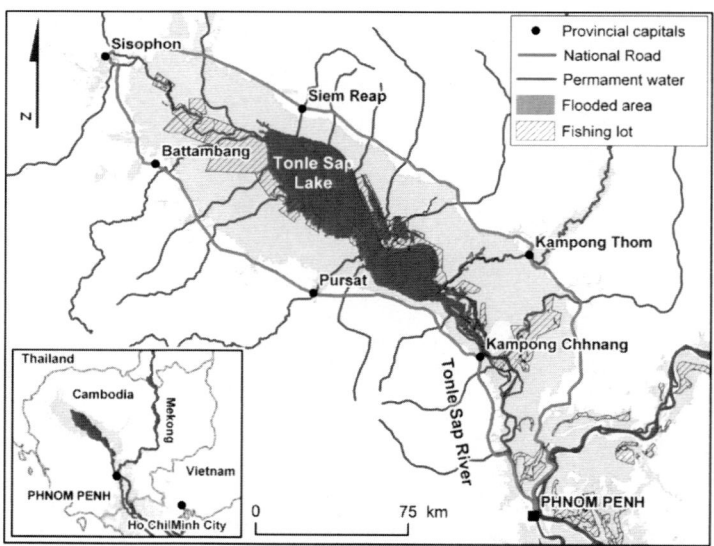

Fig. 6.2 The map of the Tonle Sap Lake area, showing the private fishing lot areas and the flooded area during exceptionally high-flood year of 2000

the Tonle Sap extends, however, much further, as it is estimated that up to half of Cambodia's population benefits from the lake's resources (Bonheur 2001).

The Tonle Sap is known for its extraordinary flood pulse system[6] with a remarkable but nevertheless rather regular seasonal variation in the lake's water volume and level (Lamberts 2006; MRCS/WUP-FIN 2007). The main driver of the flood pulse system is the Mekong River and its floods: during the wet season the water level in the Mekong mainstream rises faster than the water level in the lake. As a result, part of the floodwaters run to the Tonle Sap River, causing the river to reverse its flow back toward the Tonle Sap Lake. The lake thus loses its only outlet, and the flood-waters extend to large floodplain areas surrounding the lake. An exceptional and highly productive floodplain ecosystem has been formed based on this flood pulse system, and the Tonle Sap is considered to be among the world's most productive freshwater ecosystems and fishing grounds (Rainboth 1996; Lamberts 2001, 2006). This productivity is epitomized by the immense fish catches of the Tonle Sap Lake and the Tonle Sap River.

Taken together, the unusual flood pulse system and immense aquatic production of the Tonle Sap make it perhaps the single most vulnerable area to major changes in water quantity and quality of the Mekong River (see, e.g., Lamberts 2008; Kummu and Sarkkula 2008). The Tonle Sap is also exceptional for a lake of its size, as due to its exceptional flood pulse system, the impacts of any environmental change are felt as a combination of changes in its own basin and that of the Mekong River. The actual "impact basin" of the Tonle Sap Lake is thus not merely the lake basin (86,000 km^2), but the entire Mekong River Basin upstream from the Tonle Sap (680,000 km^2). This, naturally, makes the assessment of potential impacts to the area a particular challenge – and at the same time very much a regional issue as well.

6.5.1   Example of Impacts 1: Changing Water Levels

Different Cumulative Impact Assessment (CIA) studies have looked at the impacts of the planned hydropower development to water quantities of the Mekong (Adamson 2001; ADB 2004; DHI 2004; World Bank 2004) – and more such studies are currently being carried out (see, e.g., MRC 2009b, c, 2010). The estimates of these assessments are, however, relatively inconsistent due to the different assumptions used and the differences in the models and assessment tools themselves (Keskinen 2008; Kummu and Sarkkula 2008).

The three earlier CIAs discussed here (Adamson 2001; ADB 2004; World Bank 2004) indicate that planned development in the upper parts of the Mekong Basin

---

[6] Flood pulse is a term for an ecological paradigm integrating the processes of productivity in river-floodplain ecosystems, with a particular focus on the lateral exchange of water, nutrients, and organisms between a water body and the connected floodplain. For more information, see Junk et al. (1989).

will alter the water levels downstream and, consequently, in the floodplains. The dry season water levels are subject to rise and flood season water levels to decrease. Such changes would mean that the future flood amplitude will be smaller, leading to decreased extent of the floodplains and, consequently, to less potential spawning habitats to fish and other aquatic animals. Further, due to the smaller flood amplitude, less water will also enter to the floodplains from the mainstream.

The floodplain ecosystems need both the dry and the wet periods, and the increased low water levels would therefore permanently change the floodplain eco-systems. In the case of the Tonle Sap, the analysis of the dry season water level rise due to Mekong upstream development has in the different CIA studies been estimated as follows:

- 0.15 m increase: Estimate based on the MRC's basin-wide CIA under the Integrated Basin Flow Management (IBFM) process using the MRC's Decision Support Framework modeling tools (World Bank 2004)
- 0.30 m increase: Estimate based on the analyses of the downstream hydrological impact of the Chinese cascade of dams (DHI 2004; Adamson 2001)
- 0.60 m increase: Estimate based on the basin-wide CIA conducted within the Nam Theun 2 environmental impact assessment study using MIKE Basin model (ADB 2004)

The impact of the estimated water level rises for the dry season area of the Tonle Sap Lake is presented in Fig. 6.3.[7] The estimated rise of 0.60 m in dry season water level, as simulated by ADB (2004), would result in the permanently inundated area of 3,200 km². This would lead to the increase of the permanent lake area by nearly 1,000 km² (40%) when compared to from the current situation[8] (Kummu and Sarkkula 2008).

This kind of a rise in the lake's dry season water level, and the consequent exten-sion of the permanent lake, would result in varied impacts to the Tonle Sap and its ecosystem. Some of these impacts would be largely positive, including improved navigation possibilities due to higher water levels. The most radical impact is, how-ever, likely to be negative: increased water level would lead to permanent submer-sion of flooded gallery forest strips situated in the Tonle Sap floodplains, leading to their gradual destruction. These forest strips make an important physical barrier between the lake and the floodplain and create favorable conditions for sedimenta-tion and aquatic production. The reduction of the flooded forest area could there-fore have a significant impact on the whole Tonle Sap ecosystem and on floodplain dynamics, including the immense aquatic production of the lake-floodplain system.

---

[7] The 30-day minimum water level during the analysis period of 1997–2006 for May was 1.44 m above mean sea level (amsl), which was used as a reference level. The bottom of the lake lies at 0.6 m (amsl), and thus during the low water level, the average depth of the lake is only around 0.8 m with a lake area of around 2,300 km².

[8] Such a radical increase in the permanent lake area is explained by the fact that the Tonle Sap floodplain is extremely flat, and even small changes in the dry season water level thus permanently inundate large areas of the floodplain.

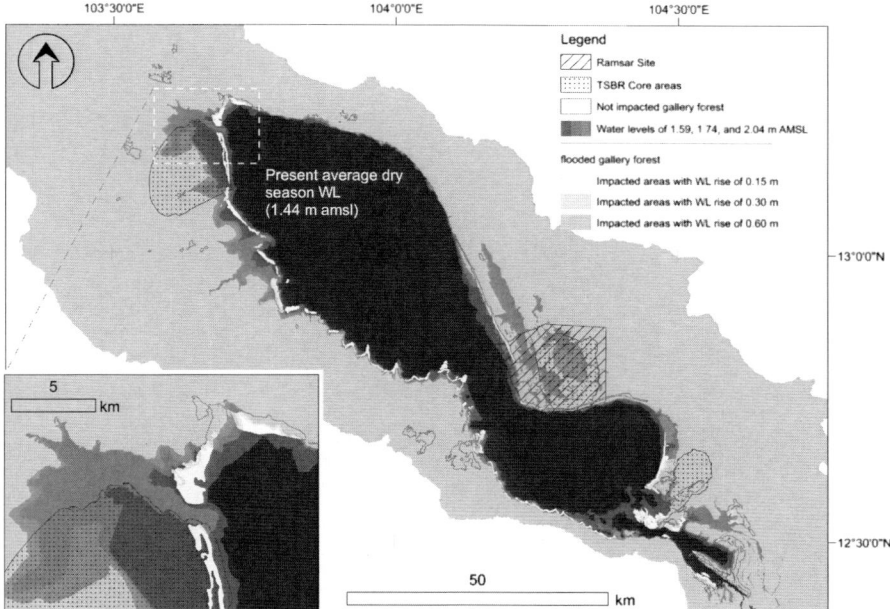

Fig. 6.3 Estimated changes in the inundated area of the Tonle Sap due to the increased dry season water level (Modified from Kummu and Sarkkula 2008)

The evolution of the biological functioning of the Tonle Sap floodplain to its present state has taken several thousands of years (Tsukawaki 1997), indicating that what is lost in the structure and productivity of the floodplain can have far-reaching and long-lasting consequences.

The CIAs also suggest that the peak water level during the rainy season would decrease, reducing the inundated area of the Tonle Sap Lake (Fig. 6.4). The total area of the Tonle Sap floodplain would therefore decrease by 7–16%, depending on the assessment used. In the case of CIA carried out by ADB (2004), the average floodplain area would decrease from present 10,750 to 9,060 km^2 by year 2025, resulting in around 15% decrease in both cumulative flooded area and flood volume. The hydropower development in the Mekong upstream would also cause changes in the flood duration in the Tonle Sap floodplain. The results from the WUP-FIN hydrological models, using input from the MRC Flow Regimes, indicate that the period of inundation would be decreasing in most parts of the floodplain by 1–2 weeks (MRCS/WUP-FIN 2007).

6.5.2   Example of Impacts 2: Changing Sediment Load

Sedimentation can be a curse or a blessing, depending on the viewpoint. For the natural environment, sedimentation is crucial, providing nutrients and other materials

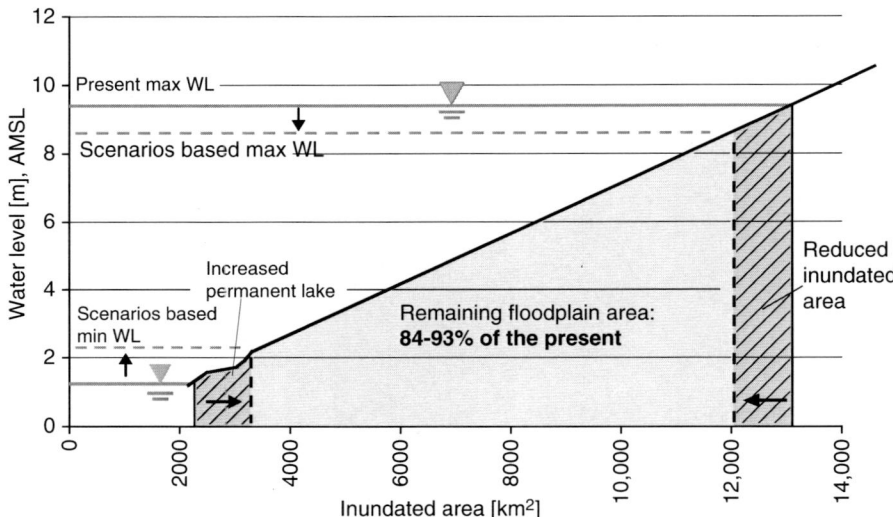

Fig. 6.4 Reduction in the area of the Tonle Sap floodplain due to the increased dry season water level (WL) and reduced wet season water level (Modified from Kummu and Sarkkula 2008)

that fuel biological productivity of the ecosystem and feed natural geomorphological processes. For humans, however, sedimentation can be problematic, causing, for example, problems for transportation and maintenance of aquatic infrastructure (Kummu et al. 2008).

The Mekong Basin yields approximately 475 km³ of water each year from a catchment area of 816,000 km² (Kummu 2008), and transports annually around $140–150 \times 10^9$ kg of total suspended sediments to the South China Sea (Milliman and Syvitski 1992). Modeled estimates for the potential sedimentation trapping for the planned cascade of eight dams in the Chinese part of the Mekong mainstream provided a result of over 90% theoretical trapping efficiency (TE) of the suspended sediment[9] (Fig. 6.5) (Kummu and Varis 2007; Kummu et al. 2010). Already this is likely to have significant impact on the whole Mekong sediment budget, as more than half of the total sediment flux originates from China (Kummu and Varis 2007; Walling 2008). The basin-wide trapping efficiency is estimated to be over 60%, if the currently planned reservoirs will be constructed (Kummu et al. 2010).

The planned construction of large-scale hydropower dams and related reservoirs will thus affect the sedimentation and erosion processes in the downstream river channels and the connected floodplains, the delta, and the coastal areas. Overall, the geomorphological impacts of the dams include bed scour, armoring of the channel,

---

[9] Theoretical trapping efficiency (TE) stands for the ratio of sediment deposition in the reservoir and synchronous total sediment input to the reservoir.

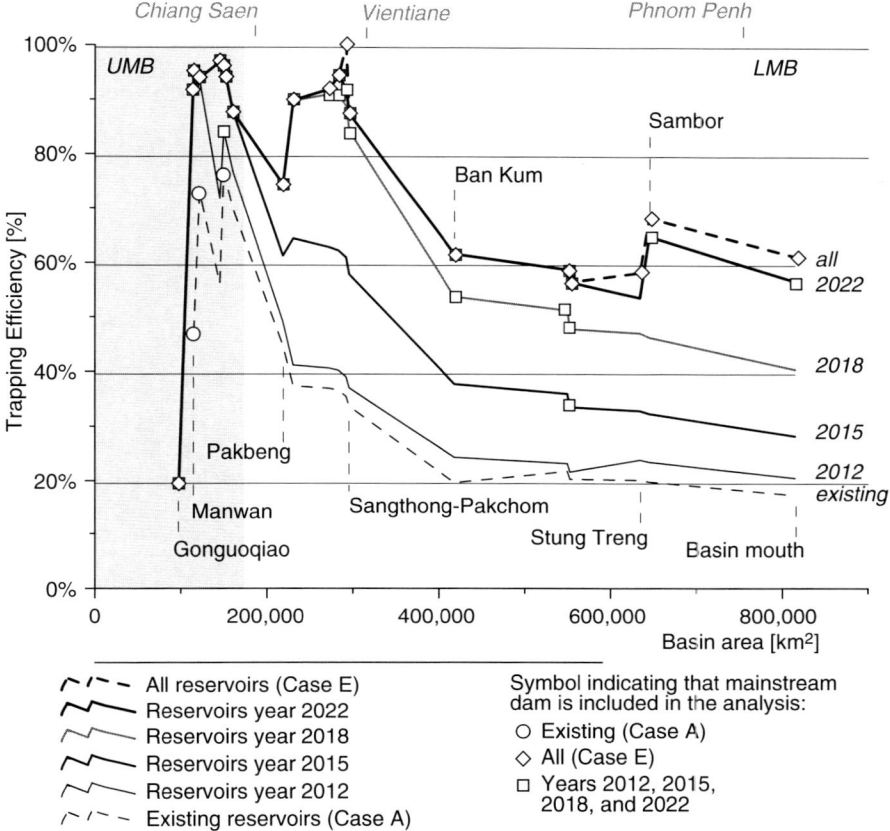

Fig. 6.5 The temporal development of the basin-wide trapping efficiency (TE) for each of the existing and planned mainstream dam locations and at the basin mouth (area = 816,000 km²) (Modified from Kummu et al. 2010)

bar and island erosion, and channel degradation and narrowing. Impacts on floodplain ecosystems are more difficult to predict, but riffles and pools are likely to be eroded. In addition, the reduced suspended sediment concentration in the floodwaters is likely to have impact on both aquatic and agricultural production as the amount of nutrients flowing to the floodplains system gets reduced (Kummu and Varis 2007).

In the case of the Tonle Sap system, the nutrients bound to suspended sediments are considered important for the system to maintain its long-term sustainability and high productivity (Kummu et al. 2008). Consequently, changes in the sediment load of the Mekong River will have a direct impact on the sediment load from the Mekong River to Tonle Sap Lake, and therefore most probably also on the high aquatic production of the lake.

6.5.3   Example of Impacts 3: Changing Ecosystem Productivity

The Tonle Sap flood pulse is largely (52%) driven by the water that is pushed up into the lake by the reversed flow of the Tonle Sap River during the rise of Mekong River flood (Kummu and Sarkkula 2008). As discussed above, the Mekong flood-waters do not only bring water, but also nutrient-laden sediments which are mostly deposited into the floodplain. The floodwaters integrate the terrestrial vegetation into the aquatic phase of the ecosystem, and this interaction forms the driving force for the high ecosystem productivity of the lake. Very little is known, however, about the exact relation between ecosystem productivity and the flood pulse (MRCS/WUP-FIN 2007).

The scenario work carried out within the WUP-FIN project estimated the cumulative impacts of the changing floodplain conditions in the Tonle Sap. The focus of the assessment was on the changes caused by so-called Flow Regime 3[10] of the MRC that was the most intensive water resources development scenario at the time (MRC 2006a). The simulation results for the Tonle Sap can be summarized as follows:

- The inundated floodplain habitat would be reduced by around 15%.
- The period of inundation would be shortened by 1–2 weeks.
- The increased dry season water level would inundate permanently a major part of the flooded forest around the lake, thus extending the permanent lake area.
- Dissolved oxygen conditions would worsen by extending strongly anoxic period in the floodplain during early flooding due to slowly rising flood.
- Sediment and nutrient input to the lake with the floodwaters would be reduced.

While providing sound estimates on the potential changes in floodplain produc-tivity is particularly challenging due to complex nature of the Tonle Sap system (Lamberts and Koponen 2008), initial estimates of the cumulative impact of the changes in above-mentioned factors were made as well. The cumulative impacts were estimated by introducing a cumulative indicator for floodplain productivity potential by giving an estimate for the minimum and maximum value for each individual factor. These estimates gave a value in the order of 25% reduction in the floodplain productivity potential, even with rather conservative estimates for indi-vidual indicator changes.[11] Although the linkages between the primary production,

---

[10] The Integrated Basin Flow Management process of the MRC assessed the impacts of three different Flow Regimes that were compositions of numerous characteristics of the hydrological system under concern. The Flow Regime 3 included most intensive development of the three regimes, including an approximate 4.5 times increase in hydropower electricity production and a 40% increase in irrigated area (MRC 2006a).

[11] This estimate is consistent with the assessment made by the expert panel within the Phase 2 of the MRC's Integrated Basin Flow Management (IBFM) process in 2006. The panel estimated that the Flow Regime 3 would result at least in an overall 20–30% reduction in the productivity potential of the Tonle Sap Lake and its floodplain (MRC 2006a).

fish production, and the fish catches are very complex, it can be assumed that any loss of primary production will directly result in the loss of secondary production and, consequently, in the reduction of the fish catches[12] (MRCS/WUP-FIN 2007). Due to the remarkable significance of the Tonle Sap's fisheries for Cambodia and even for the entire Mekong system, this kind of reduction would have severe consequences both economically and socially.

6.6 Multiple Crossroads: Water Development in a Transboundary Setting

The examples presented above point out some of the physical and environmental impacts that hydropower development in the Mekong Basin is likely to have, including potentially radical changes in water quantity and quality as well as in the ecosystem productivity. Yet, the examples represent only some of the potential impacts, and the actual overall impact to Mekong ecosystem will naturally be a combination of the different impacts. These combined impacts vary across different spatial and temporal scales (Kummu 2008), and also extend to broader issues than just hydrological and environmental impacts. In the case of Tonle Sap, for example, it remains difficult to provide reliable estimates on what would be the actual cumulative impacts of different basin development scenarios due to the complexity of the Tonle Sap system and weak understanding of the main drivers for the lake's high aquatic productivity.

The studies presented in this chapter provide thus an example of the major challenge related to current impact assessment practices in the Mekong: the problem of assessing comprehensively the cumulative impacts of basin development options. Cumulative assessment is particularly challenging in the case of complex systems such as the fisheries or floodplain dynamics, including the Tonle Sap system (Keskinen 2008; Kummu 2008). A meaningful impact assessment of crosscutting issues such as these would require a holistic approach that utilizes and integrates expertise from several disciplines and makes use of a number of different models and impact assessment frameworks. Yet, many of the existing impact assessment approaches have a relatively narrow focus, and they thus tend to "compartmentalize" the environment and social systems into selected indicators and sectors only[13] (Lamberts 2006; Keskinen 2008).

Consequently, despite enormous resources put into the different water management and impact assessment processes, they have not been that efficient in capturing comprehensively the combined impacts of different development plans at different

---

[12] For more discussion on primary productivity, please see MRCS/WUP-FIN (2007) and Lamberts and Koponen (2008).

[13] See also MRC (2009b) with its separate – although closely connected – assessments of hydrological, environmental, social, economic, and fish-related impacts.

parts and levels of the Mekong River Basin. As argued by Keskinen (2008), the reasons for this can be found from broader challenges with current basin-wide impact assessments, including:

- Reliability and representativeness of the information used in the assessments
- Challenge in addressing the different spatial and temporal scales
- Problems in assessing the crosscutting impacts
- Lack of true public engagement in the assessment processes

Despite these challenges, the results from the assessments are used to guide development planning in the basin. For example, the World Bank and the Asian Development Bank (ADB) stated in relation to their Mekong Water Resources Assistance Strategy that "the analytical work on [MRC] development scenarios has, for the first time, provided evidence that there remains considerable potential for development of the Mekong water resources" (World Bank and Asian Development Bank 2006). This statement has been criticized by different actors (Middleton 2007; IUCN and IWMI 2007b), and it is also much bolder than the more careful interpretation given by the modelers themselves (Käkönen and Hirsch 2009).[14] For this reason, the statement has also been used as an example of the use of the impact assessment results to justify certain kind of decisions and policies.

Indeed, the discussion about water development and its impacts is closely related to the differing valuations and understandings of the river and the resources it provides. Currently, the concept of "balanced development" (see, e.g., MRC 2006b; World Bank and Asian Development Bank 2006) seems for the key regional players such as the World Bank and the ADB as well as for the MRC and its member countries' governments to be closely connected to centralized, large-scale interventions such as hydropower dams. However, this kind of view tends to neglect the fact that the basin's waters have already for centuries been used and livelihoods developed through a diverse small-scale use of water-dependent resources, most importantly fish and wetlands. "Balanced development" can therefore become a euphemism that is used to hide vested interests promoting certain kinds of development paths and paradigms.

Consequently, to use the analogy applied in this chapter, the crossroads seem to be strong between centralized water development, on the one hand, and alternative approach with emphasis on more local-level development and better consideration of existing livelihood sources, on the other. Most current development plans focus on relatively large-scale, technocratic interventions that support irrigated agriculture, water diversions, and hydropower. Yet, a majority of the population in the basin relies on livelihoods that are smaller-scale and more dependent on natural resources. Worryingly, the planned large-scale developments are in many cases undermining

---

[14] It is also interesting to note that the World Bank and the ADB were selective on what they consider relevant from the MRC's findings: the strategy builds much more on the MRC's hydrological modeling exercises than on the MRC's fisheries studies that would not allow such a straightforward statement.

these more traditional livelihoods by impacting negatively the availability of and access to common pool resources such as fish (Phillips et al. 2006; MRCS/WUP-FIN 2007; Keskinen 2008).

The differing views and valuations also impact the ways different management and assessment methods are considered. The national and regional organizations focusing on large-scale utilization of the river resources tend to promote general, centralized frameworks and approaches for water management and impact assessment. For example, the World Bank and the ADB relied on the centralized Decision Support Framework (DSF) of the MRC in their Mekong Water Resources Assistance Strategy, while the MRC itself recently adopted an IWRM approach to support the implementation of the DSF (MRC 2006b, 2009a). While the IWRM as a theoretical concept puts together several well-intentioned ideas and objectives, it has also been criticized to easily lead to centralized, predefined management practices that lack proper understanding of local contexts and remain largely technical and even mechanical processes (see, e.g., Biswas 2005; Warner et al. 2008; Keskinen 2010).[15]

Such criticism relates to the general vagueness of the IWRM concept that has been seen to make IWRM prone to misinterpretations and even intentional misuse (Molle 2008). As noted by Svendsen et al. (2005), IWRM has also a strong normative content: it implicitly suggests that social, environmental, and economic aspects are compatible, when they in reality are often – including the current setting in the Mekong – in contrast with each other, making the entire IWRM procedure a highly political process. Overall, the challenges of current management and assessment practices have led to suggestions that impact assessment and management activities should build on different kinds of assessments at different levels, instead of prefixed and too narrowly defined management and assessment frameworks (see, e.g., Cash 2000; Cash and Moser 2000).

This thus brings us to the second emerging crossroads in the Mekong: that of different methods, approaches, and tools. While different "commonly accepted" standard approaches are promoted as means to achieve balanced development, their actual implementation remains only partially successful. One of their main challenges is that their use leads easily to the neglect of local contexts and needs, and that they fail to understand the fundamental differences between different spatial and temporal levels (see, e.g., Keskinen 2008, 2010; Warner et al. 2008; Käkönen and Hirsch 2009). Consequently, complementary approaches and frameworks for management are being discussed and suggested, many of them highlighting the need for more diverse approaches making simultaneous use of various methods.

The good news is that there already exists a variety of impact assessment processes in the Mekong Basin, implemented by governmental agencies, regional

---

[15] The MRC does acknowledge the challenges related to the IWRM approach, noting that "It is recognized that there is no blueprint for achieving IWRM and that various management instruments, enabling environments and institutional entities are involved. In this respect, the MRC does not expect IWRM to be achieved quickly, and certain aspects of IWRM are likely to remain at the national level and not be fully achieved at the basin scale" (MRC 2006b: 21–22).

organizations, as well as by the academia and the NGOs (see, e.g., MWBP and IUCN 2005; Lazarus et al. 2006; MRC 2006a; Swift 2006; MRCS/WUP-FIN 2007; ADB 2008; Bezuijen et al. 2008; TKK and SEA START RC 2009; Keskinen et al. 2010). These processes provide a remarkable knowledge base about the estimated impacts at different scales, and about the strengths and weaknesses of different assessment methods. The challenge is that they are currently neither properly coordinated nor properly embedded in the decision-making structures. Better coordination between the assessment processes and, in particular, increased interaction between the assessments at different scales would thus be potentially very beneficial for impact assessment in the basin. Particularly important would be to capture better the diverse experiences from the local-level assessments and take these as a basis for broader, regional assessments.

6.7 Conclusions: Way Forward

This chapter has discussed the estimated impacts of planned water development in the Mekong Basin, concluding that such impacts are likely to be significant, impacting both the environment and the societies along the Mekong in remarkable ways. At the same time, however, we highlighted the challenges of current assessment methods, underlined by the problems related to approaches relying on methods with too narrow focus. As a corollary, we argued for the existence and significance of two major crossroads: one related to different development paths and decisions, and the other one to the approaches and methods used to assess the potential impacts of such decisions.

6.7.1 Where to from the Crossroads? [16]

What would then be the possible ways forward from the two crossroads? First of all, we see that the distribution of estimated benefits, costs, and risks from the basin development should form an elementary starting point for any development plan. Practically all current impact assessments estimate that the planned water development in the Mekong Basin will result in remarkable, largely negative impacts to the water-related resources, most importantly fish. Consequently, there is a need for thorough discussion on whether the people in the Mekong countries are really willing to bear the consequences of losing an essential part of their ecosystem services and food security, and, overall, to accept such unequal distribution of the benefits and costs of planned water development.

---

[16] This section draws on Keskinen (2008) and Sarkkula et al. (2009).

Related to this, it needs to be recognized that most of the current assessment procedures tend to overly "scientize" and depoliticize the knowledge production,[17] and as a result, the moral and political dimensions of water resources development are at risk of being excluded from the discussions (Käkönen and Hirsch 2009). The assessments are never simply objective, technical processes, as already the definition of the issues to be looked at – and thus the relevant group of experts to analyze them – is a value-laden act. Similarly, the selection and use of assessment tools is shaped by certain assumptions, values, and power relations. No assessment tool should therefore be treated as a simple "truth machine."

Instead of hasty decisions about the way forward, few steps aside are therefore needed to properly discuss the conclusions of different impact assessment processes, and – based on this discussion – to assess the plausible development alternatives and their implications for the Mekong. As the planned development is likely to have remarkable impacts to the people in the basin, this pause should be coupled with radical strengthening of development dialogues with stakeholders in different riparian countries. The MRC's current assessments are, together with its increased emphasis on transparency and participation, promising steps forward, and should therefore be both supported and critically discussed.

Overall, we anticipate that the improved consideration of differing opinions and views on water development would bring to the fore the uneven share of benefits and risks between upstream and downstream as well as between social groups. This, in turn, is likely to draw more attention to the potential of alternative, smaller-scale development options as an alternative path complementing – and partly even replacing – the current path that tends to focus on larger-scale, centralized water development.

Following from this, the way forward for current impact assessment approaches seems much clearer: better utilization of and tolerance toward the diversity of different assessment methods and forms of knowledge. Indeed, we believe that water management and related impact assessment in the Mekong River Basin would benefit from a more multiscale approach that combines assessments from lower levels up to the regional level, and makes better use of interdisciplinarity and participatory approaches. Assessments of complex environmental and social issues should also have long-term perspective, building on adaptive, learning-orientated process.

Due to complexities and uncertainties involved in the assessment of development alternatives, there is also a need to use a set of different models and assessment approaches – instead of just a single model or approach – when assessing the

---

[17] By "scientization" we mean that science is given an instrumental and decisive role in legitimating policy (Bäckstrand 2004). The expectation that political consensus about development plans can simply follow from scientific consensus of the impacts and consequent trade-offs is, however, rather problematic. Scientization can hinder debate about the different development policies, and instead lead to narrower discussions about the scientific validity of the estimated impacts or to the issues of remediation and mitigation (cf. Szerszynski 1996; Wynne 2002; Demeritt 2006).

impacts to the water-related ecosystems and livelihoods.[18] While such a diversity of differing assessment methods causes increased divergence between the estimates, it also enhances the credibility and transparency of the results and decreases the possibility of completely unrealistic estimates. The variation in change estimates can also be beneficial in communicating the inherent uncertainties related to such estimates, facilitating discussion about the ways the impact assessment should actually be used.

6.7.2   The Other Side of the Coin: The Political Aspects of Water Management

These kinds of practical recommendations provide, however, only one side of water resources management. While acknowledging and appreciating such recommendations, it is crucial to recognize the highly political nature of water development, and consequently, of related planning and impact assessment processes. The underlying reasons – and solutions – for the present-day challenges with water resources development are therefore likely to lie beyond merely methodological issues, and can instead be found from broader political processes related to water management both within and between the riparian countries.

The political nature of water management has examples in the Mekong Region as well. The importance of Mekong fisheries, for instance, has been accepted widely by the riparian governments and regional organizations, but the results from the fisheries studies have been used only selectively when arguing for certain development policies (Friend et al. 2009; Käkönen and Hirsch 2009). It is also important to note that the political relations between the Mekong countries are largely building on growing economic cooperation and regional integration, with regional water management decisions being often subjugated to these broader processes.[19] This is vividly exemplified by the meager role that the Mekong River Commission (MRC) has had – despite its theoretically strong mandate – on shaping the development paths in the Mekong[20] (see, e.g., Keskinen 2006; Dore and Lazarus 2009).

---

[18] Global climate models present a good example of the benefits of this kind of multimodel use: The Intergovernmental Panel on Climate Change (IPCC) uses results from over 20 different climate models when producing their change estimates for global climate change, making use of the diversity of differing methods and increasing the credibility of their estimates.

[19] The growing regional cooperation does not, however, necessarily mean that the countries would be giving up their sovereignty on making water-related decisions. In fact, as argued by Fox and Sneddon (2007), the Mekong Agreement can even be seen to promote "environmental securitization" of the riparian states.

[20] It must also be remembered that the MRC members include only four Lower Mekong countries, with China – an increasingly influential regional hegemony – being outside the organization and its decision-making processes. See also discussion about hydro-hegemony by Zeitoun and Warner (2006).

Even if the certainty of the impact estimates increases, it is therefore by no means self-evident that the increased knowledge will have an effect on the actual decision making about water management and development. In order for alternative development options to gain ground, the politics behind seemingly science-based decisions need to be brought into debate as well (Käkönen and Hirsch 2009). Different water management and impact assessment methods are often used in overly consensual ways, resulting in situations where the politics are taken out from the decision making. The objectives of IWRM, for example, are rarely in harmony with each other, but are in fact often antagonistic (Molle 2008). As a result, IWRM can – despite its calls for a balance between economic, social, and environmental issues – be used to give legitimacy to approaches where the priority is given to economic growth, and harmful environmental and social impacts are presented as lamentable but inevitable losses that just need to be compensated.

Within this kind of frames, research findings on the severity of potential impacts do not create a consideration of alternative pathways. Instead, the entire crossroads – that of real choices, competition of ideas, popular participation, and, at the end of the day, democratic decisions – disappears and there appears to be just one possible way forward. The question is thus not anymore which development path to take, but just being how to proceed along the one that somehow already got selected.

Consequently, what is needed for real crossroads to emerge in the Mekong Region is the fostering of political dialogue at both national and regional levels. The domains of alternative visions and development values should gain more space and louder voice so that stronger articulations of differing pathways would emerge. These alternative development options should be assessed together with the options given by the current developmentalist agenda, and their benefits and costs discussed openly. Only this enables an emergence of real crossroads, where informed, deliberate decisions on different development pathways can be taken.

Acknowledgments This chapter summarizes the findings from the so-called Mekong IWRM research project that was implemented 2005–2007 by the Water & Development Research Group at the Helsinki University of Technology (now part of Aalto University). Many of the research findings have therefore been discussed also elsewhere, most importantly in the Mekong at the Crossroads special issue of Ambio (Vol. 37, No. 3). The Mekong IWRM project was funded by the Finnish Ministry for Foreign Affairs and the Academy of Finland, while the writing process for this chapter was supported by the Academy of Finland project 133748. The authors would like to thank all Mekong IWRM project counterparts and contributors for their encouragement and support. Our research has been closely linked with the Lower Mekong Modelling Project (WUP-FIN) that operated under the Mekong River Commission. Dr. Juha Sarkkula and Jorma Koponen deserve therefore special thanks for their collaboration as well as for their crucial contribution to the findings presented in this chapter. Thank you also to Joakim Öjendal, Stina Hansson, and Sofie Hellberg, as well as for an anonymous reviewer for the constructive comments.

References

Adamson PT (2001) Hydrological perspectives on the lower Mekong basin: the potential impacts of hydropower developments in Yunnan on the downstream flow regime. Int Water Power Dam Constr 53(3):16–21

ADB (2004) Cumulative impact analysis and Nam Theun 2 contributions – final report, NORPLAN and EcoLao, Asian Development Bank, Manila, Philippines

ADB (2008) Lao people's democratic republic: preparing the cumulative impact assessment for the Nam Ngum 3 hydropower project, Final CIA report – Main report, technical assistance consultant's report, Vattenfall power consultant AB in association with Ramboll Natura AB and Earth Systems Lao, Asian Development Bank (ADB), Manila

Bäckstrand K (2004) Civic science for sustainability. Glob Environ Polit 3(4):24–41

Bezuijen MR, Timmins R, Seng T (eds) (2008) Biological surveys of the Mekong river between Kratie and Stung Treng Towns, Northeast Cambodia, 2006–2007. WWF Greater Mekong – Cambodia Country Programme, Cambodia Fisheries Administration and Cambodia Forestry Administration, Phnom Penh

Biggs D (2004) Between rivers and tides: a hydraulic history of the Mekong delta, 1820–1975. PhD thesis, University of Washington, Seattle

Biswas AK (2005) Integrated water resources management: a reassessment – a water forum contribution. In: Biswas AK, Varis O, Tortajada C (eds) Integrated water resources management of south and south-east Asia. Oxford University Press, New Delhi, pp 319–336

Bonheur N (2001) Tonle Sap ecosystem and value. Technical Coordination Unit for Tonle Sap, Ministry of Environment, Phnom Penh

Cash DW (2000) Distributed assessment systems: An emerging paradigm of research, assessment and decision-making for environmental change, viewpoint. Glob Environ Chang 10(4):241–244

Cash DW, Moser SC (2000) Linking global and local scales: Designing dynamic assessment and management processes. Glob Environ Chang 10(2):109–120

Demeritt D (2006) Science studies, climate change and the prospects for constructivist critique. Econ Soc 35(3):453–479

DHI (2004) Study on natural reverse flow in the Tonle Sap river. Danish Hydraulic Institute (DHI) Water & Environment/Mekong River Commission Secretariat, Vientiane

Dore J, Lazarus K (2009) De-marginalizing the Mekong river commission. In: Molle F, Foran T, Käkönen M (eds) Contested waterscapes in the Mekong region – hydropower, livelihoods and governance. Earthscan, London, pp 357–381

Dugan P (2008) Mainstream dams as barriers to fish migration: international learning and implications for the Mekong. Mekong river commission. Catch Cult 14(3):9–15

Fox CA, Sneddon C (2007) Transboundary river basin agreements in the Mekong and Zambezi basins: Enhancing environmental security or securitizing the environment? Int Environ Agreements Polit Law Econ 7(3):237–261

Friend R, Arthur R, Keskinen M (2009) Songs of the doomed: the continuing neglect of capture fisheries in hydropower development in the Mekong. In: Molle F, Foran T, Käkönen M (eds) Contested waterscapes in the Mekong region – hydropower, livelihoods and governance. Earthscan, London, pp 307–331

Hirsch P, Jensen KM, Boer B, Carrard N, FitzGerald S, Lyster R (2006) National interests and transboundary water governance in the Mekong. Australian Mekong Resource Centre at The University of Sydney in collaboration with Danish International Development Assistance, Sydney

IUCN, TEI, IWMI, M-POWER (2007a) Exploring water futures together: Mekong region waters dialogue. Report from regional dialogue. The World Conservation Union (IUCN), Thailand Environment Institute (TEI), International Water Management Institute (IWMI) & M-POWER, Vientiane, Lao PDR

IUCN, TEI, IWMI, M-POWER (2007b) Feedback on WB/ADB joint working paper on future directions for water resources management in the Mekong river basin. Mekong Water Resources Assistance Strategy (MWRAS), the World Conservation Union (IUCN), Thailand Environment Institute (TEI), International Water Management Institute (IWMI) and M-POWER, Vientiane, Lao PDR

Jakeman A, Letcher R, Norton J (2006) Ten iterative steps in development and evaluation of environmental models. Environ Model Softw 21(5):602–614

Junk WJ, Bayley PB, Sparks RE (1989) The flood pulse concept in river-floodplain systems. In: Dodge DP (ed) Proceedings of the international large river symposium (LARS), vol 106. Canadian Special Publication of Fisheries and Aquatic Science, Ottawa, pp 110–127

Käkönen M (2008) Mekong delta at the crossroads: more control or adaptation? Ambio 37(3):205–212. Available online at: http://water.tkk.fi/global/publications

Käkönen M, Hirsch P (2009) The anti-politics of Mekong knowledge production. In: Molle F, Foran T, Käkönen M (eds) Contested waterscapes in the Mekong region – hydropower, livelihoods and governance. Earthscan, London, pp 333–365

Keskinen M (2006) The lake with floating villages: socioeconomic analysis of the Tonle Sap lake. Int J Water Resour Dev 22(3):463–480

Keskinen M (2008) Water resources development and impact assessment in the Mekong basin: which way to go? Ambio 37(3):193–198. Available online at: http://water.tkk.fi/global/publications

Keskinen M (2010) Bringing back the common sense? Integrated approaches in water management: lessons learnt from the Mekong. Dissertation for the degree of doctor of science in technology, Water and development publications, Aalto University, Helsinki, Finland

Keskinen M, Käkönen M, Tola P, Varis O (2007) The Tonle Sap lake, Cambodia; water-related conflicts with abundance of water. Econ Peace Secur J 2(2):49–59

Keskinen M, Chinvanno S, Kummu M, Nuorteva P, Snidvongs A, Varis O, Västilä K (2010) Climate change and water resources in the Lower Mekong River Basin: putting adaptation into the context. J Water Clim 1(2):103–117

King P, Bird J, Haas L (2007) The current status of environmental criteria for hydropower development in the Mekong region: a literature compilation. Consultants Report to Asian Development Bank, Mekong River Commission Secretariat and World Wide Fund for Nature, Vientiane

Kummu M (2008) Spatio-temporal scales of hydrological impact assessment in large river basins: case Mekong. Doctoral thesis, Water and development publications. Helsinki University of Technology, Espoo, Finland. Available online at: http://water.tkk.fi/global/publications

Kummu M, Sarkkula J (2008) The impact of Mekong river flow alteration on the Tonle Sap flood pulse and flooded forest. Ambio 37(3):185–192. Available online at: http://water.tkk.fi/global/publications

Kummu M, Varis O (2007) Sediment-related impacts due to upstream reservoir trapping in the lower Mekong river. Geomorphology 85:275–293

Kummu M, Penny D, Sarkkula J, Koponen J (2008) Sediment: curse or blessing for Tonle Sap lake? Ambio 37(3):158–163. Available online at: http://water.tkk.fi/global/publications

Kummu M, Lu XX, Wang JJ, Varis O (2010) Basin-wide sediment trapping efficiency of emerging reservoirs along the Mekong. Geomorphology 119(3–4):181–197

Lamberts D (2001) Tonle Sap fisheries: a case study on floodplain gillnet fisheries. Asia-Pacific Fishery Commission, Food and Agricultural Organisation (FAO), Bangkok

Lamberts D (2006) The Tonle Sap lake as a productive ecosystem. Int J Water Resour Dev 22(3):481–495

Lamberts D (2008) Little impact, much damage; the consequences of Mekong River flow altera-tions for the Tonle Sap ecosystem. In: Kummu M, Keskinen M, Varis O (eds) Modern myths of the Mekong – a critical review of water and development concepts, principles and policies. Water & Development Publications, Helsinki University of Technology, Espoo, Finland, pp 3–18. Available online at: http://www.water.tkk.fi/global/publications

Lamberts D, Koponen J (2008) Flood-pulse alterations and productivity of the Tonle Sap ecosystem: a model for impact assessment. Ambio 37(3):178–184

Lazarus K, Dubeau P, Bambaradeniya C, Friend R, Sylavong L (2006) An uncertain future: biodiversity and livelihoods along the Mekong river in northern Lao PDR. IUCN, Bangkok/Gland

Middleton C (2007) The ADB/WB/MRC 'Mekong water resources assistance strategy': justifying large water infrastructure with transboundary impacts. Paper prepared for critical transitions in the Mekong region, Regional Center for Sustainable Development, Chiang Mai, 29–31 Jan 2007

Miller F (2006) Environmental risk in water resources management in the Mekong delta: a multiscale analysis. In: Tvedt T, Jakobsson E (eds) A history of water: water control and river biographies, vol 1. Tauris, London, pp 172–193

Milliman JD, Syvitski JPM (1992) Geomorphic/tectonic control of sediment discharge to the ocean: the importance of small mountainous rivers. J Geol 100:525–544

Mirumachi N, Nakayama M (2007) Improving methodologies for transboundary impact assessment in transboundary watercourses: navigation channel improvement project of the Lancang-Mekong river from China-Myanmar boundary marker 243 to Ban Houei Sai of Laos. Int J Water Resour Dev 23(3):411–425

Molle F (2008) Nirvana concepts, narratives and policy models: insight from the water sector. Water Altern 1(1):131–156

MRC (Mekong River Commission) (2005) Overview of the hydrology of the Mekong basin. Mekong River Commission, Vientiane

MRC (Mekong River Commission) (2006a) Flow regime assessment. Integrated basin flow management report no 8: flow-regime assessment. Water utilization program/environment program. Mekong River Commission, Vientiane, Lao PDR

MRC (Mekong River Commission) (2006b) Strategic plan 2006–2010: meeting the needs, keeping the balance. Mekong River Commission, Vientiane

MRC (Mekong River Commission) (2008a) Existing, under construction and planned/proposed hydropower projects in the lower Mekong basin, Sept 2008. Map produced by the Mekong River Commission. Available online at: http://www.mrcmekong.org/programmes/hydropower.htm. Accessed 07 July 2009

MRC (Mekong River Commission) (2008b) Fish migration emerges as key issue at regional hydropower conference. Mekong river commission. Catch Cult 14(3):4–8

MRC (Mekong River Commission) (2009a) IWRM-based basin development strategy for the lower Mekong basin. Incomplete consultation draft no 1 – Oct 2009. Mekong River Commission, Vientiane

MRC (Mekong River Commission) (2009b) Assessment methodology – economic, environmental and social impact assessment of basin-wide water resources development scenarios. Technical note, draft – Oct 2009. Basin Development Plan Programme, phase 2, Mekong River Commission, Vientiane

MRC (Mekong River Commission) (2009c) Inception report. MRC SEA for hydropower on the Mekong mainstream. Mekong River Commission (MRC) and International Center for Environmental Management (ICEM), Vientiane

MRC (Mekong River Commission) (2010) Synthesis of initial findings from assessments – assessment of basin-wide development scenarios. Technical note 1, draft – Feb 2010. Basin Development Plan Programme, phase 2, Mekong River Commission, Vientiane

MRCS/WUP-FIN (2007) Final report – part 2: research findings and recommendations. WUP-FIN phase 2 – hydrological, environmental and socio-economic modelling tools for the lower Mekong basin impact assessment. Mekong River Commission and Finnish Environment Institute Consultancy Consortium, Vientiane, Lao PDR. Available on-line at: http://www.eia.fi/wup-fin/wup-fin2/publications.htm

MWBP, IUCN (2005) Thai Baan research on the ecology and local history of the seasonally-flooded forest in the lower Songkhram river basin. The Mekong Wetlands Biodiversity

Conservation and Sustainable Use Programme (MWBP) and the World Conservation Union (IUCN)

Öjendal J (2000) Sharing the good: modes of managing water resources in the lower Mekong river basin. PhD dissertation, Department of Peace and Development Research, Göteborg University, Gothenburg, Sweden

Phillips D, Daoudy M, McCaffrey S, Öjendal J, Turton A (2006) Trans-boundary water cooperation as a tool for conflict prevention and for broader benefit-sharing. Phillips Robinson and Associates, Windhoek

Rainboth WJ (1996) Fishes of the Cambodian Mekong. FAO Species Identification Field Guide for Fishery Purposes/Food and Agriculture Organisation of the United Nations (FAO), Rome

Rayner S (2003) Democracy in the age of assessment: reflections on the roles of expertise and democracy in public-sector decision making. Sci Public Policy 30(3):163–170

Sarkkula J, Keskinen M, Koponen J, Kummu M, Nikula J, Varis O, Virtanen M (2007) Mathematical modelling in integrated management of water resources – magical tool, mathematical toy or something in between. In: Lebel L, Dore J, Daniel R, Koma YS (eds) Democratizing water governance in the Mekong region. Mekong Press, Chiang Mai, pp 127–156

Sarkkula J, Keskinen M, Koponen J, Kummu M, Richey J, Varis O (2009) Mekong hydropower and fisheries – what are the impacts? In: Molle F, Foran T, Käkönen M (eds) Contested waterscapes in the Mekong region – hydropower, livelihoods and governance. Earthscan, London, pp 227–249

Svendsen M, Wester P, Molle F (2005) Managing river basins: an institutional perspective. In: Svendsen M (ed) Irrigation and river basin management: options for governance and institutions. CABI Publishing, Wallingford, pp 1–18

Swift P (2006) Livelihoods in the Srepok river basin in Cambodia: a baseline survey. The NGO Forum on Cambodia, Phnom Penh

Szerszynski B (1996) On knowing what to do: environmentalism and the modern problematic. In: Szerszynski B, Lash S, Wynne B (eds) Risk, environment & modernity: towards a new ecology. Sage, London

TKK, SEA START RC (2009) Water and climate change in the lower Mekong basin: diagnosis and recommendations for adaptation. Water and Development Research Group, Helsinki University of Technology (TKK), Finland and Southeast Asia Regional Center (SEA START RC), Chulalongkorn University, Thailand. Water & Development Publications, Helsinki University of Technology, Espoo, Finland

Tsukawaki S (1997) Lithological features of cored sediments from the northern part of the Tonle Sap lake, Cambodia. The international conference on stratigraphy and tectonic evolution of southeast Asia and the South Pacific, Bangkok, Thailand, 19–24 Aug 1997, pp 232–239

Varis O, Kummu M, Keskinen M (2008) Mekong at the crossroads. Ambio 37(3):146–149. Available online at: http://water.tkk.fi/global/publications

Vörösmarty CJ, Sahagian D (2000) Anthropogenic disturbance of the terrestrial water cycle. Bioscience 50(9):753–765

Walling D (2008) The changing sediment load of the Mekong river. Ambio 37(3):150–157

Warner J, Wester P, Bolding A (2008) Going with the flow: river basins as the natural units for water management? Water Policy 10(Suppl 2):121–138

WEC (2007) 2007 Survey of energy resources. World Energy Council (WEC), London

World Bank (2004) Modelled observations on development scenarios in the lower Mekong basin. Mekong regional water resources assistance strategy (MWRAS), Prepared for the World Bank with MRC cooperation, Technical assessment by Geoff Podger and Richard Beecham and review, observations and conclusions by Don Blackmore, Chris Perry and Robyn Stein. World Bank, Vientiane, Lao PDR

World Bank, Asian Development Bank (2006) WB/ADB Joint working paper on future directions for water resources management in the Mekong river basin. Mekong water resources assistance strategy (MWRAS), June 2006. The World Bank and Asian Development Bank, Washington, DC

Wyatt AB, Baird IG (2007) Transboundary impact assessment in the Sesan river basin: the case of the Yali falls Dam. Int J Water Resour Dev 23(3):427–442

Wynne B (2002) Risk and environment as legitimatory discourses of technology: reflexivity inside Out? Curr Sociol 50(3):459–477

Zeitoun M, Warner J (2006) Hydro-hegemony – a framework for analysis of trans-boundary water conflicts. Water Policy 8(5):435–460

Chapter 7
Negotiating Flows in the Mekong

Kate Lazarus, David J.H. Blake, John Dore, Worawan Sukraroek, and David S. Hall

Abstract Negotiating water flows should be an essential part of river basin management in the Mekong Region. Environmental flows, or E-flows, can be an important tool to assist. Central to E-flows is the recognition that ecosystems not only have their own intrinsic value, but also provide humans with essential goods and services. E-flows work concentrates on establishing water flow regimes which recognise ecosystem needs whilst trying to satisfy social and economic demands. Putting E-flows theory into practice requires the integration of a range of disciplines including engineering, law, ecology, economy, hydrology, sociology, political science and communication. This chapter investigates use of E-flows approaches in the Mekong Region. The first case explores the establishment of national constituencies for E-flows. Three subsequent cases have experimented with E-flows to aid river basin management negotiations where, as is usual, there are competing

K. Lazarus (✉)
Water Governance Specialist and Mekong MSP Coordinator, M-POWER and the Challenge
Program on Water and Food, Vientiane, Lao PDR
e-mail: katelazarus2008@gmail.com

D.J.H. Blake
School of International Development, Faculty of Social Sciences, University of East Anglia,
NR4 7TJ, Norwich, United Kingdom
e-mail: djhblake@yahoo.co.uk

J. Dore
Senior Water Resources Advisor – Mekong Region, Australian Agency for International
Development (AusAID), Vientiane, Lao PDR
e-mail: johndore@loxinfo.co.th

W. Sukraroek
School of Geosciences, University of Sydney, Madsen Building (F09), NSW 2006, Australia
e-mail: s_warawan@yahoo.com

D.S. Hall
Mekong Sub-region Social Research Centre, Ubon Ratchatani University,
Ubon Ratchathani, Thailand

J. Öjendal et al. (eds.), *Politics and Development in a Transboundary Watershed:*
The Case of the Lower Mekong Basin, DOI 10.1007/978-94-007-0476-3_7,
© Springer Science+Business Media B.V. 2012

positions, interests, priorities and development objectives. The results from the case studies indicate that the E-flows approach has the potential to contribute more to decision making for improved water governance in the Mekong Region.

7.1 Introduction

Negotiating water flows should be an essential part of river basin management in the Mekong Region. Environmental flows, or E-flows, can be an important tool to assist diverse basin stakeholders in the process of negotiation and better understanding of the resource. A widely used definition is 'An environmental flow is the water regime provided within a river, wetland or coastal zone to maintain ecosystems and their benefits where there are competing water uses and where flows are regulated' (Dyson et al. 2003). To take an E-flows approach[1] requires establishing water flow regimes which recognise ecosystem needs whilst trying to satisfy social and economic demands. E-flows can be a useful approach to facilitate participatory exploration and negotiation and can lead to more informed decision making on water resource issues within countries and across boundaries. As river resources become increasingly subject to competition, a multi-stakeholder approach to E-flows management is suggested that requires and enables the integration of a range of disciplines, including engineering, law, ecology, economics, hydrology, sociology, political science, environmental science, fisheries science and communications.

Whilst water resource management at the national or river basin level is typically a top-down, government-led and controlled process, E-flows approaches provide methods to bring together various stakeholders to determine an equitable way to share the bounty of rivers without over-exploiting or damaging flow-dependent ecosystems. With the pace of major infrastructure developments accelerating in the Mekong Region, E-flows approaches can assist in bringing different perspectives and actors to the table to discuss water allocation decisions.

E-flows pilot studies have been implemented in Vietnam and Thailand, led by the International Union for Conservation of Nature (IUCN) in partnership with provincial governments, International Water Management Institute (IWMI) and others. Mekong River Commission (MRC) has also experimented with E-flows, under the banner of Integrated Basin Flow Management (IBFM). These experiments, whilst presented under the rubric of E-flows, have had different conceptual foundations leading to some confusion about terminology and the way the E-flows approach can be applied. For example, whilst the MRC set out to 'apply the principle of E-flows and its concepts to determine appropriate water resources and development options and the maintenance of flows', the formal governance procedure adopted by the MRC focused on using E-flows to determine minimum flows (MRC 2006a, b). In contrast, IUCN's work has aimed to use E-flows in deliberative processes to negotiate acceptable flows between competing uses.

---

[1] There are many different methodologies, depending on the chosen emphasis. For a recent review see Tharme (2003).

In this chapter, we reflect on the different terminology used for E-flows and its linkage to Integrated Water Resources Management (IWRM). This is followed by a discussion on the importance of water for development and ecosystems. We then look at how E-flows have been used in the Mekong Region by reviewing four cases (see Fig. 7.1). These are (1) a working group process in six countries, led by IUCN, which produced a translated E-flows text in six languages of the region; (2) a rapid E-flows assessment in Vietnam's Huong River Basin; (3) the IBFM project of the MRC in the Lower Mekong Basin; and (4) a multi-stakeholder, interdisciplinary[2] E-flows assessment in Thailand's Nam Songkhram River Basin. Throughout and in our concluding remarks, we explore the limitations and potential for E-flows as a constructive negotiation tool.

This chapter is based on the direct experience of the authors in the implementation of these E-flows projects. Two authors were involved in designing, conceptualising and coordinating of IUCN's activities around E-flows. Two others led the social impact assessment of the MRC's IBFM. Another author was closely involved in the planning, coordination and implementation of the Nam Songkhram River E-flows pilot project in Thailand.

7.2   E-flows and Integrated Water Resources Management

The Mekong Region is faced with significant obstacles and challenges in implementing sustainable water resources management. With the overall population expected to grow significantly, and with consumption trends projected to rise in line with economic growth and growing urbanisation, freshwater ecosystems will be placed under increasing pressure. River flows will be impacted by increased withdrawals for irrigated agriculture, the construction and operation of hydropower dams, abstraction for domestic and industrial water supply to large cities, widespread wetland encroachment and degradation and navigational improvement projects to improve trade between countries. These demand-induced changes in flow will be compounded by future climate change pressures (WWF 2009).

E-flows is an approach to assist in meeting the above-mentioned challenges. E-flows challenge the way in which rivers have been managed in the past.

'Historically water has been managed from a supply perspective with an emphasis on maximizing short-term economic growth from the use of water. Little thought has been

---

[2] We draw on definitions for interdisciplinary and multidisciplinary approaches from Klein and Newell (1997). 'Interdisciplinary activities link together and integrate information and methodologies from two or more separate, traditional, and/or artificial disciplines. Such linkages and integration creates a multi-faceted picture of a topic through the exploration and synthesisation of various approaches/views. To ensure success, these intersections and connections among disciplines require a non-hierarchical sharing of intellectual authority and continuous dialogue among ALL participants'. 'Multidisciplinary offers present information and methodologies on a given topic from more than one separate, traditional, and/or artificial discipline without linking or integrating them. Typically, this approach presents different disciplinary approaches to the same topic through the juxtaposition of terminology, methodology, assumptions, and goals. It addresses a topic by presenting disciplines as "stand alone" or parallel views'.

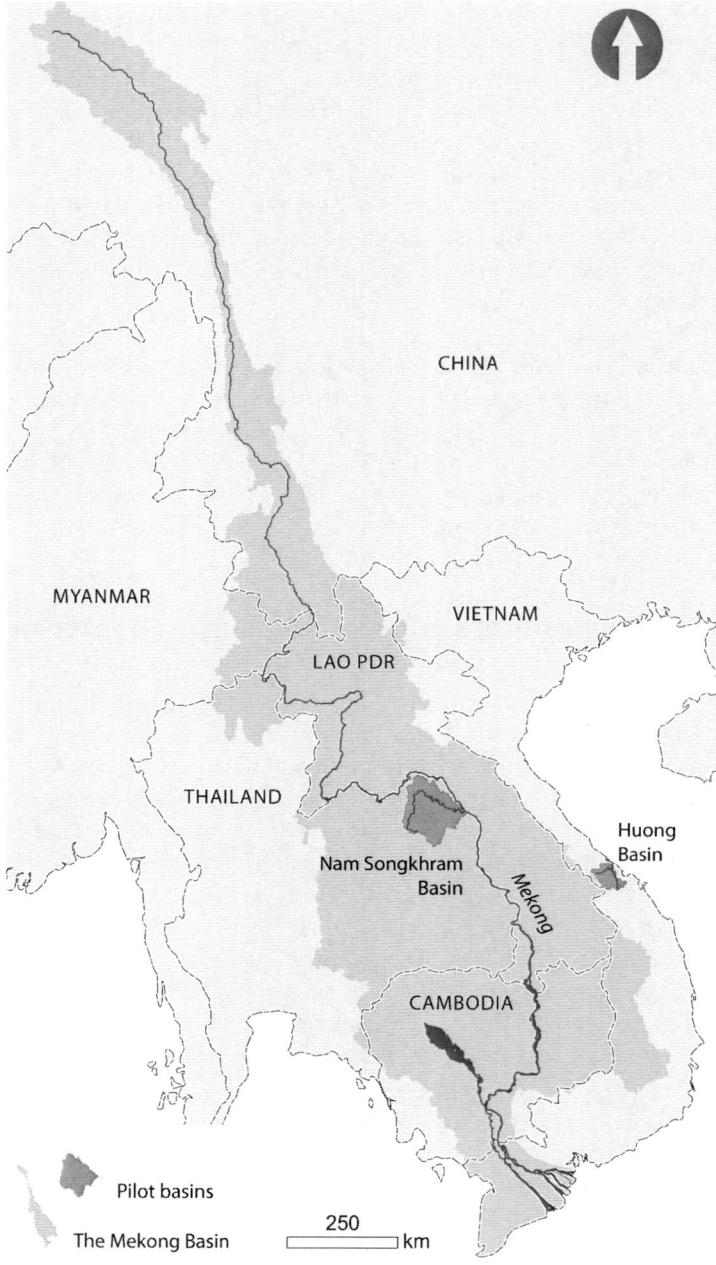

Fig. 7.1 Pilot basins with E-flows experimentation in the Mekong region (Source: Kummu 2010)

Table 7.1 Levels of environmental flows assessments (EFA): implications

Method	Resources	Time	Confidence	Resolution	Status
Desktop – rapid	Low	2 days – 2 weeks	Low	Low	Planning guide
Intermediate	Medium	~8 weeks	Medium	Medium	Preliminary EFA
Comprehensive	High	~32 weeks	Medium/high	Medium/high	Full EFA

given to the health of the resource itself and there is poor understanding of the implications of overuse or declining river health. Water resource managers are now trying to come to terms with the need to take a more holistic view of the river system using the Integrated Water Resource Management (IWRM) paradigm. They increasingly understand that one needs to take care of aquatic ecosystems and the resources they provide for long-term economic viability' (Dyson et al. 2003).

E-flows are an integral part of holistic IWRM, which is generally understood as 'a process that promotes the coordinated development and management of water, land and related resources, in order to maximise the resultant economic and social welfare in an equitable manner without compromising the sustainability of vital ecosystems' (Global Water Partnership 2000).

IWRM can also be seen as 'a political procedure that aims for sustainability of use; a process of balancing all water demands and supplies including those for environmental maintenance; an iterative approach that recognises the need for adaptive management; and a way of life' (King and Brown 2009). E-flows can create the necessary space for dialogue and negotiations between diverse stakeholders, an essential part of the 'political procedures' referred to by King and Brown. E-flows can also provide the necessary technical information required for flow allocation decisions.

The E-flows approach recognises the environment as an important sector in its own right. By identifying the environment as a sector, it encourages decision makers to identify ways of balancing demand for irrigation, hydropower, domestic and industrial purposes, also known as addressing the 'triple-bottom' line of environmental, social and economic development. In some countries, such as South Africa, legislation protects the rights of the environment to a water allocation, compelling planners to determine the flows required to sustain ecosystems through an E-flows assessment.

During the last five decades, about 200 different approaches/methods have been described as E-flows and more than 30 countries have begun to use such assessments (see Table 7.1 for levels of assessment) in the management of water resources (Tharme 1996; Arthington and Zalucki 1998; Tharme 2003; King et al. 2003). E-flows application varies nationally, regionally and globally. In many cases, E-flows is not well understood and terminology used differs markedly. For example, other terms used instead of E-flows include minimum flow,[3] instream

---

[3] 'Minimum flows are used to describe the retention of enough flow to maintain river connectivity, especially for fish passage, but this is usually only one component of the flow regime that needs to be maintained, and there are few instances where an environmental flow consists of just a minimum flow' (Hirji and Davis 2009).

flow,[4] ecological flow and environmental demand, and these may have a different meaning to each individual or organisation (Moore 2004; Hirji and Davis 2009). These terms are not always considered synonymous with E-flows and may not be equivalent, but they are certainly in the same family of approaches, grounded in an appreciation of ecosystem values.

There is evidence of a broad adoption of E-flows around the world, with developed countries leading the way and many developing countries with advanced interest (Moore 2004). For example, Tharme and Smakhtin (2003) reported increasing research and practice in E-flows assessments within a number of Asia's developed and developing countries including Indonesia, Japan, Korea, Nepal, Pakistan, Sri Lanka and Taiwan.

The definition we presented for E-flows at the beginning of this chapter has been further elaborated by the Brisbane Declaration[5] as '*the goal of environmental flow management is to preserve socially valued freshwater ecosystem benefits and biodiversity*'. This Declaration calls on governments, development banks and water managers to take immediate action to recognise the benefits of environmental flows in water resources planning and to apply the important lessons learned from efforts to implement environmental flows around the world (Brisbane River Symposium 2007).

Central to the E-flows concept is the recognition that ecosystems not only have their own intrinsic value, but also provide humans with essential services (Environmental Flows Network 2004). E-flows are important for freshwater dependent ecosystems, as these require a share of water to maintain their function. Freshwater ecosystems provide a wealth of food, fuel, medicine and fibre, water purification, aquatic organisms and wildlife habitat, tourism and recreational opportunities, navigation, employment and opportunities for culture and spiritual renewal (Krchnak 2006).

E-flows can be an important approach for strengthening the relationship between functioning healthy ecosystems and sustainable livelihoods for both urban and rural communities. Ecosystems are not only a user of water in competition with other users, but the base from which socially valued resources are derived and supported, and without which no sustainable uses are possible.

Ultimately, E-flows should inform negotiations on how much flow from the river in question should be allocated to different sectors. As allocations to a given sector are likely to reduce the amount or timing that can be allocated to another, the process is likely to be contentious and involve 'trade-offs' that may not be acceptable to all parties. Because win-win solutions are rare, the key role of E-flows is to make costs and benefits of flow allocations to different sectors explicit so that meaningful and informed negotiations can take place.

---

[4] 'Instream flows imply the flows needed to maintain ecosystem services from flows within the river channel, but this excludes the often important floodplain flows that overtop the channel' (Hirji and Davis 2009).

[5] The Brisbane Declaration presents principles and a global action agenda that responds to the most urgent needs to protect rivers globally. It calls for action that strongly encourages the governments, development banks and water managers to take immediate action to recognise the benefits of environmental flows in water resources planning and implement the important lessons learnt from efforts to implement environmental flow protection around the world. Implementation of environmental flow provisions as part of effective water governance is seen as integral (Brisbane River Symposium 2007).

The concept of trade-offs is controversial in the Mekong Region (Friend and Blake 2009). Here, the emerging tensions between the hydropower and capture fishery sectors are significant, and there are concerns that 'trade-offs' may be presented as a technical solution to decisions that will ultimately be highly political in nature because of their differential impacts on rural and urban populations (with the former being expected to make sacrifices for the supposed benefit of the latter). A further concern is that a focus on the impacts of trade-offs *between* sectors, such as hydropower and fisheries, will detract from a deeper consideration of alternatives *within* a sector (e.g. what alternatives exist in the energy sector and how are risks and benefits shared for each alternative?).

One important dimension of E-flows is that the approach entails an analysis of how changed flow regimes ultimately impact people. However, predicting the likely magnitude of social consequences of changed river flows is complex as detailed information on the extent of people's use of the impacted river goods and services is first needed. Nevertheless, the argument in this chapter is that E-flows can be a powerful tool to contribute to water resources decision making and negotiating water allocation.

7.3 Water for Development and Ecosystems

An important aspect of E-flows is that it is often said to involve a compromise between 'water for development' and 'water for nature' (Environmental Flows Network 2006). This is particularly relevant and has significant resonance with the key tenets of the 1995 Mekong Agreement, which emphasises the balance between sustainable development and economic growth. Article 5 of the 1995 Mekong Agreement focuses on the principle of reasonable and equitable utilisation. During negotiations of the Agreement, it became clear that the countries wanted a flexible agreement that could adapt to future conditions but still be specific enough to ensure that their interests were met on each one's priority concerns (Radosevich in Dore et al. 2010).

With the global population quadrupling in the past century, and much of this occurring within the Asia-Pacific Region, it is unavoidable that areas of irrigated agricultural land will continue to expand in some locations, and water withdrawals from freshwater ecosystems will continue to increase (Richter et al. 2006). Ongoing infrastructure development continues to compromise the health of freshwater ecosystems in myriad ways.

Failing to allocate enough water for the environment is likely to cause the ecosystems already stressed to deteriorate, thereby seriously affecting local livelihoods (IWMI 2005). It is important to dispel the myth that water allocated to the environment is water unavailable for humans (Krchnak 2006). Moreover, it needs to be recognised that the maintenance or rehabilitation of river systems is linked to poverty alleviation (Environmental Flows Network 2006). Where it is the case, it is necessary to demonstrate the dependency of livelihoods on ecosystem goods and services. This is already occurring in the Mekong Region. For example, Osborne (2007) discusses the rapid changes that are taking place in international rivers of Southeast Asia, such as the

Salween and Mekong Rivers, with evidence to suggest that environmental and social impacts of development are no longer fringe issues in the region.

The challenge now is to move beyond the false dichotomy of the water for development or nature debate and better appreciate interdependencies. Using an interdisciplinary E-flows approach to assist in negotiating water resources decision making may be one way of improving the governance of water resources management in the Mekong Region.

7.4   Integrating and Negotiating

There are a number of challenges and opportunities in applying the concept of E-flows to water management. Krchnak (2006) argues that 'a challenge that applies to both developed and developing countries is the complexity of developing E-flows recommendations that are aligned with social goals, particularly in ways that involve all stakeholders in deciding upon the health of a country's rivers'. To have any prospect of wide acceptance, diverse stakeholder groups need to be meaningfully engaged in the development of an E-flows regime. E-flows advocates stress 'putting E-flows into practice is not easy' (Dyson et al. 2003). It is emphasised that E-flows requires the integration of a range of disciplines across the social, political and natural sciences, whilst facilitating a learning process between various stakeholders that attempts to understand and bridge their different and often competing interests over water. Multi-stakeholder platforms (MSPs) can be an important part of the flows negotiation process. Dore (2010) argues that 'Multi-stakeholder platforms are a part of governance in which different stakeholders are identified and, usually through representatives, invited and assisted to interact in a deliberative forum that focuses on: sharing knowledge and perspectives; generating and examining options; and informing and shaping negotiations and decisions'.

It is recognised that there can be no single best method, approach or framework to determine E-flows. What works in one river basin, institutional or socio-cultural setting, may not necessarily work in another. This may depend on the intended objectives of developing E-flows, resulting in different stakeholders preferring one method over another. Hence, it is important to be aware of and attentive to lessons from past experiences.

7.5   E-flows in the Mekong Region

The countries in the Mekong Region are at varying levels of E-flows understanding or implementation (see Table 7.2). As pointed out above, the 1995 Mekong Agreement emphasises the principle of reasonable and equitable utilisation of the waters in the Mekong River. Radosevich in Dore et al. (2010), additionally points out some of the underlying interests of the four Lower Mekong Basin countries

Table 7.2 Status of E-flows application in the Mekong region

Cambodia	Participated in the MRC-led Integrated Basin Flow Management (IBFM) study for the Lower Mekong River Basin. E-flows have not been mainstreamed into national river basin management, national legislation or policies
China	E-flows are currently under experimentation within the Yellow River and Suzhou River with respective objectives for flushing sediment and water quality. Environmental flows assessments have been undertaken in the Zhangzi River and Tarim Basin. The regulations of the Yellow River Conservancy Commission note that allocation should include integrated management of domestic, agricultural, industrial and environmental water use (see www. yrcc.gov.cn). The Nature Conservancy (TNC) is assessing water flows in the Yangtze River Basin to identify ways to sustain river ecosystems and improve the design and operation of dams in order to minimise impacts on the river and its fish populations. WWF is also engaged in E-flows activities in the Yangtze. There is also emerging research in hydrological engineering and aquatic ecology relevant to E-flows in China. However, there is no clear legislation or policies at a national level where E-flows have been explicitly recognised
Lao PDR	Lao PDR was part of the IBFM study for the Lower Mekong River Basin undertaken by the MRC. E-flows have not been mainstreamed into river basin management, national legislation or policies at this stage. At the 1st National Waters Dialogue[a] held in 2007, participants agreed that principles of E-flows should be included in the revision of the Law on Water and Water Resources
Myanmar	No evidence of use of E-flows in the country
Thailand	The use of E-flows is a relatively new tool in Thailand. The IBFM study for the Mekong River Basin included Thailand. IUCN and partners piloted an E-flows study in the Nam Songkhram River Basin in 2007. There is a legislative and policy framework in which environmental considerations need to be considered for major development. However, constraints to the official adoption of E-flows in Thailand include the complexity of land and water use, institutional structures and social diversity and availability of suitable methodologies, information and data
Vietnam	The use of E-flows as a tool for IWRM is more advanced in Vietnam than in other countries in the Mekong Region. The Huong River is an example of where an E-flows assessment has been undertaken. Vietnam was also part of the IBFM study for the Lower Mekong River Basin by MRC. The Law on Water Resources (1998) and the National Water Resources Strategy: Towards the Year 2020 recognises that to protect aquatic ecosystems, attention to ensuring E-flows, within a suite of other measures, is necessary. For example, the strategy says 'the focus on economic development and low awareness of the importance of aquatic ecosystems in the balance of nature had led to severe degradation of aquatic ecosystems, especially freshwater ecosystems, where many species are becoming rare, and some are on the brink of extinction. The case of this problem is the lack of attention to aspects such as the importance of ensuring environmental flows; the importance of water ecosystem protection when physical structures on rivers are built; and the importance of controlling the exploiting and use of water to sustainable levels' (Ministry of Natural Resources and Environment 2006)

[a]In 2007, IUCN and partners convened a national dialogue with the Government of Lao PDR and other stakeholders to discuss key water-related governance concepts, including E-flows.

(Cambodia, Lao PDR, Thailand and Vietnam) around the time of negotiations for the 1995 Mekong Agreement.

> Thailand, initially concerned that other countries might try to veto proposed developments, advanced the position that each riparian should unilaterally be able to use tributary waters within its territory without approval of the other riparians. Vietnam, in agreement with Cambodia and Laos, was very concerned about maintaining flow levels in the mainstream during the dry season and advanced the position that the use of water from the mainstream should be agreed upon by a joint technical committee before any water was diverted.... The countries tentatively agreed in principle on the requirements for water use during the wet season in the tributaries and the mainstream, pending resolution of conditions on water use from the mainstream during the dry season. Article 6 of the Agreement details requirements for maintaining base flows on the mainstream including the dry season..... The package the countries eventually accepted builds on differences in the location (tributary or mainstream), kind of use (inter- or intra-basin), timing (wet and dry season) and type of procedural requirements (ranging from notification to prior consultation to specific agreement).....

Tharme (2003) indicated that by the early 2000s, China, Lao PDR, Thailand and Vietnam had expressed interest or were in the early stages of undertaking E-flow assessments.[6] Vietnam has explicitly included E-flows in national legislation and policies. Many decisions in the Mekong Region have to be made about flow management. This has become increasingly vital as major decisions are underway, for example, about whether to build up to 12 hydropower projects along the mainstream of the Mekong River south of China. Negotiating E-flows could provide a conceptual framework and information base to aid this critical decision making, but the approach is not yet widely understood or used in the region. E-flows could be a part of the initial key decision making, and not just a patch for subsequently dealing with the consequences.

Factors that are limiting the adoption of E-flows in the Mekong Region include lack of awareness and understanding of the concepts associated with E-flows, lack of political support and a lack of readiness to accept and face up to complexity and uncertainty when making decisions about rivers.

Elsewhere, E-flows adoption and implementation has been particularly strong where national legislation and policies place E-flows as a requirement within an IWRM framework and integrate E-flows into natural resource management plans at the catchment scale. To be included in policies or regulations, some scientists assert that there needs to be a strong agreement on the quantitative methods for E-flows. Others argue that the social dimensions are equally important.

Currently there is limited capacity and demand to use E-flows as a negotiating tool amongst a diverse set of stakeholders in the Mekong Region, as some

---

[6]For the four lower Mekong countries, Cambodia, Laos, Thailand and Vietnam, their assessments have been largely linked to the IBFM process of the MRC. China has moved further ahead through piloting E-flows assessments in various river basins and Vietnam has piloted a rapid E-flows assessment in the Huong River Basin and incorporated minimum flows into their national strategy (see Ministry of Natural Resources and Environment 2006).

governments in the region believe they have more pressing water management issues that require immediate attention such as floods and droughts (often presented as 'crises narratives') or developing river systems that are presently 'undeveloped'. For example, there is a strong perception amongst some senior Thai government planners and policy makers that river water flowing to the sea unutilized for irrigation or hydropower is being 'wasted' (Molle et al. 2009).

The Mekong Region (see Fig. 7.1) is a place where E-flows work has been experimented with on the mainstream of the Mekong River, led by the MRC, as part of an IBFM process. This work had many supporters – more outside MRC than inside – who wanted it to be successful and contribute to river development scenario debates. To be useful, the work must continue to focus on key areas of concern. For example, it is widely recognised that Cambodia is heavily dependent on the productivity of the Tonle Sap Great Lake wild fishery which is threatened by ecological disruption to the natural 'flood pulse' system caused by water infrastructure development–related changes to the wet and dry season flow regimes. Clarifying the causes and extent of the threats, and then including that in basin development debates, is an important component of E-flows work. In subbasins of the Mekong, such as the Nam Songkhram River Basin – another flood pulse river system – in northeast Thailand, IUCN and partners also experimented with E-flows approaches incorporating economic, ecological, social and transboundary dimensions.

Catalysed by major floods in 1999, IUCN worked with provincial authorities and IWMI to explore E-flows in Vietnam. The Huong River Basin is a classic case of competing uses for water, competing views about whether a flood event is a disaster or a natural occurrence and a range of views about what should be done. In short, it is just the kind of situation where an E-flows approach can contribute.

Guttman (2006) referred to the need to 'recognize the value of products and services provided by the river system' and that 'the question of values is fundamental to identifying environmental flow requirements'. Most would agree. In many parts of the Mekong Region, there has been an urgent need to find a way for different perceptions of value to be expressed and heard in some type of deliberative process of exchange. At the transboundary level, recent efforts in the Mekong River Basin have included the Basin Development Plan (BDP) scenarios and a Strategic Environmental Assessment (SEA) process, specifically for the mainstream Mekong hydropower dams. The BDP and SEA processes, and whatever comes after them, would do well to revisit participatory modes of E-flows as a useful approach to ensure water-related negotiations do not become beholden to one or other sectoral or stakeholder set of interests.

These examples are elaborated in this chapter to show the different experiments with piloting E-flows in the Mekong Region and the challenges and opportunities for implementation.

7.6 Multi-stakeholder Translations in the Mekong Region: Building Capacity and a Constituency

Between 2005 and 2007, IUCN and partners led an initiative of mini multi-stakeholder processes that yielded translations of the book *Flow: The Essentials of Environmental Flows* (Dyson et al. 2003) into Burmese, Chinese, Khmer, Lao, Thai and Vietnamese languages.

The English version of the book was unpacked, the concepts debated and the book reassembled in a local language translation by teams of State and non-State actors from different sectors. This aimed to establish a wider understanding of and constituency for E-flows throughout the Mekong Region which was considered necessary before any piloting could take place.

The process and product were equally important. In practice, the process of translation unfolded differently in each country, but tried to stay true to a set of agreed principles:

- State and non-State actors
- Different disciplines
- Different perspectives
- Not privileging of any particular discipline or set of actors
- Take the time required to build a constituency
- Deliberate choice not to just have one translator but many contributors and peer reviewers

Working papers were assembled to assist professional translators in preparing an appropriate translation of *Flow* in each country. Each working group included people with different perspectives to ensure that the translation process included collective understanding and learning.

A different process took place in China, less participatory, but still resulting in a Chinese version of *Flow* that was launched in Beijing by the Ministry of Water Resources in 2006. Non-State actors were heavily involved in the reviewing of the Ministry of Water Resources draft text and had a large impact on the final text. Cambodia, Vietnam and Laos were all deliberative with State and non-State actors. The Burmese translation was undertaken by a high-quality civic organisation, and in Thailand, a small team of four persons led the work.

The product was then reviewed by experts from various backgrounds such as engineering, law, aquatic ecology, economics, hydrology, political science and geography. The strength was the structured process with a foundation document. The working groups and their peers became informed spokespersons about E-flows in their countries and also regionally. The process confirmed the necessity of building national constituencies before there is any likelihood of successful acceptance of any transboundary E-flows process in the Mekong Region. The weakness of the process was that these same groups of national 'translation nego-tiators' were not the same as those involved in piloting of E-flows at the trans-boundary level.

7.7 Rapid E-flows Assessment in the Huong River Basin, Vietnam

The People's Committee of Thua Thien Hue Province in central Vietnam wants to ensure responsible management of the Huong River Basin, which takes account of the health of the ecosystem and associated social and economic benefits. It is therefore supportive of the effort to learn about environmental flows, ultimately establish an environmental flow regime, and in so doing contribute to IWRM in the Huong River Basin (IUCN Vietnam 2005).

The E-flows approach was tested in central Vietnam's Huong River Basin (Thua Thien Hue Province) where flooding and saltwater intrusion have been major concerns. In 2003–2004, a multi-partner group, including the Huong River Projects Management Board (HRPMB), IWMI, IUCN and local government agencies, identified the importance of developing an IWRM strategy in the province to maintain ecosystem integrity whilst providing social, cultural and economic benefits to the local people.

Over two-thirds of the population of Thua Thien Hue Province lives within the Huong River Basin, all of whom rely directly or indirectly on the river resources for their livelihoods and well-being. The river system also provides vital functions for many of the riparian and aquatic ecosystems supporting the rich biodiversity found in the province. The Tam Giang-Cau Hai Lagoon is one of the largest river mouths of its kind in Asia and is an important asset to the local people. Flooding in the rainy season and saltwater intrusion in the dry season are major concerns in this area due to geographical and meteorological conditions (IUCN Vietnam 2005).

In order to address the concerns in the Huong River Basin and determine a multi-faceted and integrated solution to competing water uses, a rapid E-flows assessment (EFA) was initiated. The Huong River Basin was chosen as a pilot project for Vietnam largely because, whilst its problems were complex, its politics were relatively simple: the entire river flows through only one province.

Public participation is a process. It will take time, but it won't come at once. But be assured that we at the provincial government are creating an opportunity for the public to be included in the decision-making process, and the environmental flows process is one way to do this (Quote by Nguyen Ngoc Thien, Vice Chairman, Hue Provincial People's Committee in 2007 to IUCN).

A key objective of the work was to assist local water managers and users to undertake the principles and practice of E-flows as a normal part of IWRM and to build local capacity of partners to undertake such work in order to improve water resources decision making.

The results from the Huong River offered few insights from the biological and social sciences, as it was heavily focused on hydrological aspects. The main methodological focus of the Huong River were the EFA workshops, held in 2004 and 2005 to open dialogue of perceived future impacts of dams on downstream ecosystems and communities. The focus of the workshops was strongly on identifying present river conditions including river classification and hydrological, ecological and social conditions of the river basin in general and of the assessment site in particular. It was the hydrological regime that was further elaborated, identifying and distinguishing between different key elements of the flow regime (such as timing of wet and dry months and size and frequency of flood events) and their importance to ecosystems. An alternative hydrological regime scenario was estimated. A number of indicators were agreed upon and a synthesis of expert opinions of all participants into a single ecology matrix was made to demonstrate the impact of the agreed flow scenario on the various indicators. The matrix was intended to provide a tool for decision makers to weigh the various consequences of their management decisions (IUCN Vietnam 2005).

Whilst valuable skills were obtained by the stakeholders involved in the rapid EFA, more importantly there was a greater appreciation of the range of disciplines, interests and perspectives required to inform infrastructure planning and flow negotiations. However, there was still a general lack of understanding of the link between hydrology and ecology in the assessment. This was deemed as one of the common limitations of rapid methodologies and also a function of limited data. This downscaling of scope (from full to rapid assessment) could be seen as leading to a downscaling of expectations and a reduced constituency. It was a challenge given this was the first time such an assessment had been carried out and the group tended to lean towards addressing the hydrological outcomes; once the broader ecological and societal needs had been incorporated, it was felt that the hydrological results were downplayed. Thus, a start to negotiating the different and contradicting interests was carried out, but further steps are needed in the Huong for E-flows to reach its potential.

Lessons learned from the Huong River EFA were manifold. First, significant time and resources are required for implementation of an intermediate or comprehensive E-flows assessment (see Table 7.1). Second, rapid E-flows assessments require substantial reliance on expert judgement as there are always data gaps. Finally, expertise from a wide range of fields is essential to ensure a holistic approach.

Recommendations from the EFA included the need to start with open discussions amongst all stakeholders to ensure ownership of the process. Practical conditions of different regions/countries must be considered, and due attention should be paid to single-province river basins of special value, such as the Huong River Basin (IUCN 2005 and IUCN Vietnam 2005).

7.8 E-flows and the Integrated Basin Flow Management Process in the Mekong River Basin

An additional level of complexity in water resource sharing and exploitation is added when dealing with large transboundary river systems, such as the Mekong River Basin. This case example is quite different from the others explored earlier in the chapter as it is a regional case study as opposed to a 'country' study and has strong, multi-country political aspects. The different context explored here has varied implications for the potential of using an E-flows approach for negotiation.

In order to fulfil Article 6 (as mentioned above), the MRC embarked on its IBFM[7] project from 2005 to 2008 using what it called the Mekong Method.[8] This method was based loosely on the holistic approach in DRIFT (Downstream Response to Imposed Flow Transition). DRIFT is a scenario-based framework, providing decision makers with a number of options of future flow regimes for a river of concern, together with the consequences for the condition of the river (Dyson et al. 2003). The IBFM aimed to use the DRIFT approach to assess the likely biophysical, social and economic impacts based on different flow scenarios.

The overall objective of the MRC IBFM was to 'introduce an holistic, multidisciplinary approach to assess river flows from the perspective of beneficial uses (economic, social and environment) enabling discussion between the member-states on trade-offs and finally agreement on an acceptable flows regime (minimum flows) framework for basin development and flow monitoring' (Guttman 2006).

IBFM was implemented as a set of activities aimed at providing information and knowledge to decision makers on economic benefits and environmental and social impacts of development as related to changes in the flow regime and facilitate the trade-off process (MRC 2008).

IBFM was carried out by establishing a multidisciplinary team of specialists mainly formed of hydrologists, economists, ecologists and modellers to identify linkages between flow regimes, reflecting future options for water resources development in the Lower Mekong Basin (LMB), and the status of natural resources and the local communities dependent on the river and its floodplain. Subject experts were selected from the four LMB countries, supplemented with international experts to develop capacity to undertake an E-flows assessment.

Hydrological data of the Mekong River, maintained within a database by the MRC, was used to outline development scenarios and predict impacts. All of the flow regime scenarios considered different combinations and levels of possible irrigation and hydroelectric power developments in the basin. Other combinations of possible development activities were also expected to be considered (Guttman 2006).

According to Guttman (2006), 'the practical experience around the world of applying environmental flows assessments have mainly been on smaller systems

---

[7] In 2006 the MRC Council approved the *Procedures for the Maintenance of Flows on the Mainstream*. The agreement includes the following: 'Specifically, except in the cases of historically severe droughts and/or floods, the Procedures apply to the cooperation in the maintenance of flows on the mainstream at selected stations: (a) of not less than the acceptable minimum monthly natural flow during each month of the dry season under Article 6A; (b) to enable the acceptable natural reverse flow of the Tonle Sap to take place during the wet season under Article 6B; and (c) to prevent average daily peak flows greater than what naturally occur on the average during the flood season attributed to intentional water releases from manmade activities and other facilities under Article 6 C. The flows to be maintained at specified locations as stipulated in a-c above are set out in a separate document entitled Technical Guidelines to be adopted/established by the MRC Joint Committee' (MRC 2006a).

[8] The Mekong Method aimed to incorporate useful aspects of environmental flows assessment as well as more conventional hydrological studies.

often highly regulated, with an aim to restore some functions or values which has been lost (or were diminishing rapidly). Applying flow assessments to larger systems has often focused on restoring a specific component, such as salmon fisheries, [as in the case of the Columbia River in America]. The application of a comprehensive and holistic assessment of larger systems, which is still in a relatively un-modified condition (such as the Mekong River) is unusual and in the context of Asia unique'. Like most E-flows experiments, IBFM was a complicated undertaking and aimed to establish in-depth understanding and knowledge of flows and their relationship to ecosystems and people's livelihoods.

Two directions were envisaged, first, to support river basin planning by allowing different water resources development options to be assessed and provide information on costs and benefits of that development and associated impacts, and second, to contribute to the maintenance of flows on the mainstream of the Mekong (as laid out in the 1995 Mekong Agreement).

The IBFM Mekong Method developed by the MRC involved determining:

- The current hydrology of the river based on existing information such as historical flow data. This was to understand the Mekong flow season and its importance to the ecosystem and seasonal use of the Mekong resources for sustaining livelihoods
- Historical parameters to describe the flow conditions of the river and its relationship with the flow change
- Parameters to describe environmental and socioeconomic benefits and impacts
- Flow response relationships quantifying how possible future changes in hydrological parameters would probably be expected to cause changes in environmental and socioeconomic parameters

The key IBFM activities included three components: biophysical, social and economic assessment based on the DRIFT methodology. The first component encompassed a 1-year hydrological assessment of the Lower Mekong Basin culminating in the publication *Overview of the Hydrology of the Mekong Basin* (MRC 2005). The publication acknowledged the importance of four flow seasons including the role of the flood pulse in sustaining the Mekong waters ecosystems, related water resources and people's livelihoods. This provided the basis for further analysis of the flow into components and zones, which could be analysed separately with respect to flow changes (Guttman 2006). The implications of the hydrological changes were then interpreted by seven specialists with expertise in vegetation, water quality, geomorphology, modelling, fisheries economics and social science.

The second component was the social assessment which estimated (a) the number of people living within 'resource use corridors' along the length of river, (b) their likely degree of dependence on aquatic and riverine resources and (c) the likely vulnerability to changes in the abundance of the resources (MRC 2008). The first task was achieved through the use of GIS technology and available demographic data, whilst the other two used secondary data based on an extensive literature review. The social component of IBFM laid the foundation for a proper household survey that detailed the extent of dependence on fish, other aquatic animals and plants from specific ecosystems carried out in 2008/2009 under the MRC Social Impact Monitoring and Vulnerability Assessment (SIMVA) programme.

The third component consisted of economic assessments whereby all development scenarios were assessed in terms of benefits and costs. The second and the third components suggested preliminary findings with respect to the assessment of flow regimes and impacts.

The term 'trade-offs' was used throughout the IBFM study; however, it was clear that the countries were not ready to address the issues raised. The findings from MRC's IBFM Report #8 (2006b) were clear: Thailand would have the opportunity to obtain more energy whilst Laos would increase its income from the sale of hydropower. Cambodia would be most severely impacted due to the loss of the fisheries. In the context of the IBFM, the trade-offs were based on the knowledge of the three streams of assessment. A paper by Friend and Blake (2009) highlights the potential risks and contradictions resulting from an overly narrow focus by Mekong River Basin and water resources planners on the notion of trade-offs, especially with regards to the inherent tensions between the hydropower and fishery sectors. More informed decisions are possible if trade-offs are explored. Baran and Myschowoda (2009) observed that 'only with more rigorous quantitative assessments that take into account all possible impacts of dam construction will well-informed decisions be possible in the context of national and regional development plans'. And, of course, exploring trade-offs only gets you so far, eventually value judgements and priorities should also figure in decision making.

The IBFM demonstrated the DRIFT methodology in the Mekong. Three streams of study were undertaken – biophysical, social and economic assessments – with an overall goal to provide knowledge on the costs and benefits of different water resources development scenarios in the LMB. However, there was no negotiation component in this E-flows experiment. Deliberations that could have informed negotiations about 'development space'[9] were not permitted. The MRC IBFM project ended at the point of presentation of scenarios and initial results from the three streams of assessment (King and Brown 2009). Stakeholder consultation was one of the main activities designed for the IBFM project where all initial results of the assessments were to be discussed and debated in multi-stakeholder forums, supported by IBFM results. However, public disclosure of the IBFM results was withheld for more than 2 years, and multi-stakeholder consultations were never undertaken. Unfortunately, none of the publications resulting from the IBFM studies were formally published and released in the public domain. IBFM Report #8, which documented the initial assessment of the three components, was eventually released but was never formally published. There was insufficient consensus on the IBFM scenario 'results'. There were conflicting views from the riparian country representatives about the economic parameters and values of water-related resources. The prepared scenarios were claimed

---

[9] 'The concept of Development Space, which is defined by present day conditions and the negotiated limit of ecosystem degradation as basin development proceeds' (King and Brown 2009).

to be insufficiently robust and likely to be controversial, at a time when the MRC Secretariat was at pains to avoid any such controversy. There was insufficient ownership of the IBFM findings by National Mekong Committee Secretariats, and a lack of appreciation or readiness for these scenarios to be debated, and quite possibly improved and amended. IBFM was wrapped in 2008, by which time MRC was embarking on a new scenarios process as part of the second phase of the Basin Development Programme (BDP2). This scenarios process, operating from 2008 to 2010, was remarkably devoid of social and ecological character. During the Mekong Region Waters Dialogue in July 2006, participants had recommended that 'the outputs of IBFM would become inputs to political discussions, so it was essential that there was transparency in the methods and indicators used, and that the rationale for different flow regime scenarios was clearly explained. Engagement of local communities must also be encouraged, in terms of both carrying out IBFM activities and assessing the accuracy of the results. A regular mechanism for channelling information from the public should be built into the IBFM process' (IUCN et al. 2007). With IBFM being wrapped, this did not happen. However, the recommendations did influence, to some extent, the follow-up BDP2 process.

Although IBFM produced a series of booklets designed to 'transform' complex findings into accessible information and conducted a series of workshops, these were not sufficient to overcome stakeholders concerns of the process being too complex. However, the process was partly successful in introducing the IBFM/DRIFT concepts to MRC staff and consultants and the government officers involved in the IBFM project, although the methods employed to create an understanding of the complexity of the basin and its need and multi-sectoral benefits may not have been fully understood.

Despite the IBFM ending prematurely, it succeeded in conveying a clear message that the Mekong's resources will be impacted due to flow changes from high levels of infrastructure development, notably hydropower projects. In addition, it provides a foundation for follow-on processes such as the BDP2, Strategic Environmental Assessment for the Lower Mekong mainstream dams, and the next generation of the SIMVA initiative.

Global experience and guidance on E-flows requires integration of a range of disciplines and also negotiations between stakeholders to bridge different interests that compete for use of water. The reward is an improved management regime that guarantees the longevity of the ecosystem and finds the optimal balance between the various uses (Dyson et al. 2003). There are several initial lessons to learn from the IBFM process. Whilst the IBFM managed to undertake multidisciplinary research, which resulted in sharing and debating across the disciplines amongst the consultancy team and MRCS staff, including a certain level of sharing with government agencies involved in the project through national consultations, efforts to engage a wider stakeholder base were blocked due to the critical and controversial nature of the IBFM findings. Second, the complexity of the IBFM along with the limited translation of scientific findings into accessible languages was not carried out. Finally, honing in on the urgency to determine a negotiated flow regime in the

Mekong succeeded in raising awareness but also enabled lower-level government bureaucrats to stall processes.

Moving towards negotiating decisions would be a next step after data is collected and analysed. These negotiations are and should be the domain of states, anything else would be impossible. Processes like the IBFM could inform negotiation processes, but will only succeed if there is clear demand for negotiations to take place (i.e. if the status quo is being maintained, the information gathered is unlikely to be of any consequence). In the case of IBFM, part of the reason for not being able to move forwards was the lack of agreement on the initial findings which presented contentious trade-offs. There were underlying political tensions with the IBFM due to different expectations of the results of the IBFM from the governments of the four MRC member countries, which led to lack of approval of project findings. The other follow-on activities mentioned above, the BDP2 and SEA, have already made significant improvements in the way in which the MRC involves the public (where the IBFM process patently failed). However, it is still unclear as to whether the results will contribute to determining a negotiated flow regime.

7.9 An Interdisciplinary Approach to E-flows in the Nam Songkhram River Basin, Thailand

The Nam Songkhram is the second largest river basin in Thailand's Northeast Region, known as *Isaan*, covering an area of 13,128 km². It is situated in the far northeast corner of Isaan in an area bounded to the south by the Phu Phan hill range which divides the Nam Songkhram Basin from the larger Mun-Chi River Basin and to the north and east by low sandstone hills beyond which lies the broad arch of the Mekong mainstream and Lao PDR. The Nam Songkhram River is characterised over most of its course by gentle gradients and impressive meanders, with the last few 100 km of the river flowing across a broad floodplain wetland landscape, with a wide range of permanent and temporary, neutral and artificial wetland habitats present. It is a relatively unregulated river in terms of flow discharge at present, although much small-scale water resources infrastructure exists and one large-scale irrigation project regulates flows on the Nam Oon tributary. A defining feature of the Lower Nam Songkhram River Basin (LSRB) is that it experiences a widespread natural flood across its floodplain each rainy season lasting between 2 and 4 months and has close eco-hydrological connections with the Mekong mainstream. This includes a marked annual backstopping of flow and occasional flow reversal out of the Mekong and up the Nam Songkhram River, not dissimilar to the annual Tonle Sap flow phenomenon in Cambodia, albeit on a much smaller scale.

An E-flows assessment was carried out in the Nam Songkhram River Basin in 2006–2007. This was the first time such an approach had been trialled in Thailand. It was developed based on the conviction that E-flows does not only consider the importance of river flows from a physical and ecological perspective, but is also related to socioeconomic and political factors at play in the river basin.[10] The interdisciplinary E-flows work in the Nam Songkhram River Basin was designed as a preliminary step towards the provision of improved data and practical tools for river basin and water managers at national, regional and local levels elsewhere in the Mekong Region.

The E-flows approach in the LSRB combined two core elements: (1) a stepped dialogue and consultation process with key actors and stakeholders within the basin and at a national level before and after the collection of empirical data, and (2) an intermediate EFA exercise that collected field data using a range of local, regional and international experts drawn from across a range of disciplines and institutions.

The overall emphasis of the study was to be placed on comprehending the ecological and social linkages on the floodplain wetlands with dependent communities. It was stressed that an appreciation of the flood pulse and key hydrological events such as the magnitude, duration, timing, frequency of flood and peak and low flow characteristics would be important, so these could be related to individual disciplines.

It was recognised that the E-flows approach was almost entirely new to the team and there were no local precedents to draw upon, which proved both a barrier in terms of familiarity with E-flows and opportunity to test a locally appropriate approach. The basic methodology employed is described as follows:

- Collection of data from three representative sites in the Basin, using an intermediate E-flows assessment (IFA) approach (see Table 7.1), integrating the skills and the knowledge base of a range of specialists using an interdisciplinary exercise. The field studies were timed to coincide with the two extremes of flow condition, i.e. peak river levels in late August/early September 2006 and minimum flows in March 2007. This allowed first-hand visualisation of flow variations in consecutive seasons and provided snapshots of the biophysical and socioeconomic-cultural conditions pertaining at these critical times of the annual hydrological cycle. Eight days were spent on each seasonal assessment, with two days spent at each field site.

- Following the fieldwork in 2007, some possible future development scenarios were drawn up for the Nam Songkhram Basin. Based on the field findings and individual's 'expert opinions', the team reassembled in May to make broad predictive summaries about likely impacts on flow, ecosystems and livelihoods that might result from the implementation of each possible scenario. The outcomes were used to inform proceedings of a subsequent multi-stakeholder dialogue that brought together a wide range of basin actors, including state sector representatives from agencies assigned the task of developing water resources at the basin or provincial level and local community representatives.

---

[10]The role that people play both as beneficiaries of the wider riverine ecosystem, and at the same time modifiers of the ecosystem, are key to understanding E-flows. It has been recognised that 'flow is the key driver of the system' (Dyson et al. 2003), where seasonal natural flows with alternating periods of flooding and low flows are seen to be the principle driving force behind the productivity and diversity of tropical river-floodplain systems (see Junk and Wantzen 2004).

This flexible methodological approach emerged from an iterative process of negotiation and compromise between the parties involved that was considered appropriate to the local context. It should be recognised that whilst the E-flows approach adopted may reveal political aspects of control and allocation of water resources operation in the basin, it is itself a highly political process that requires a degree of reflexivity by the actors involved. A key subgoal of the research was to build individual capacity to understand the importance of E-flows whilst working in a multidisciplinary team. Every effort was made for team members to work together and share insights. This approach, it is believed, helped to break down some of the barriers resulting from reductive or single disciplinary research approaches and fostered a better appreciation of commonalities across the societal and natural science spheres.

The Nam Songkhram River Basin E-flows study continually stressed the inter-disciplinary linkages at the core of the process that underpinned the effort. It helped cement and broaden cross-disciplinary understanding amongst the team members and allowed them to more confidently talk about issues outside their main field of knowledge when communicating with interested observers, according to partici-pants' feedback. Simply put, they began to recognise the wider linkages between flow, ecosystem and livelihoods towards the end of the process, which were not immediately apparent to the team at the start. Whilst an increased knowledge and understanding of the river floodplain system and how hydrological flows affect it was built up amongst team members, there were concerns that the process needed much longer and greater resources invested to become accepted in the key Thai state water resources management agencies. An unexpected output was the realisa-tion that there are several other analogous flows occurring on and around the flood-plain, beyond the material water flows that were the primary object of the team's attention. These included the spatial and temporal flow of natural resources on and off the floodplain, the flow of people in and out of communities or across basin and national borders and the more symbolic flows of knowledge and power associated with water as a scarce resource, which, it was felt by some team members, were equally deserving of further attention in future flows studies. Hence, and significantly, this wider definition of 'flows' opens up new and hitherto unexplored avenues of understanding relating to the human-nature interactions in the Mekong River Basin, which may hint at clashes of interest between dominant discourses of development and alternative interpretations of development priority agendas.

The field study provided an improved understanding of the close eco-hydrolog-ical relationship between the mainstream Mekong and the Lower Nam Songkhram River. Because of the primary influence of the Mekong mainstream on LSRB flood timing, duration and extent (as highlighted in WUP-FIN[11] models), any attempt to

---

[11] The WUP-FIN was a complementary project to the Mekong River Commission Water Utilisation Programme (MRC/WUP). It was funded by the Government of Finland. The project was imple-mented over two phases. Phase 1 (2001–2004): Modelling of the Flow Regime and Water Quality of the Tonle Sap and phase 2 (2004–2006): Hydrological, Environmental and Socio-Economic Modelling Tools for the Lower Mekong Basin Impact Assessment.

control flooding by building flow control infrastructure on the Lower Nam Songkhram River or main tributaries like the Nam Oon is likely to be futile and counterproductive, creating new and undesirable environmental impacts, which so far have not been taken into account in state-backed project proposals. This is evident from existing top-down attempts to build irrigation and flood control infrastructure on the mainstream, such as the redundant watergates built at Ban Nong Gaa, Ban Dung District, Udon Thani and numerous smaller tributaries elsewhere, all of which were built without social or environmental impact assessments.

The LSRB floodplain is in the advanced stages of an ecological transformation from being dominated by natural wetland habitat diversity to a more simplified ecosystem with fewer habitats and less biodiversity. This is principally as a result of wholesale removal of natural vegetation and conversion to agricultural land, in particular paddy fields, cash crops and recently, eucalyptus and rubber plantations. The ecological impacts of this transformation are not well studied, but abundant anecdotal and some empirical evidence collected during this study and others suggests that they are serious in terms of biodiversity loss and reduced aquatic productivity (Blake et al. 2009). The loss of ecosystem functions and services appears to be having serious negative impacts on fishery productivity and local livelihoods through food and income security declines, reflected in such phenomena as increasing labour migration out of the area, reversing an earlier trend of local in-migration during the 1960–1990 period. These macro-trends suggest a loss of floodplain agro-ecosystem resilience (see Vidal et al. 2010) and that there is a concomitant social as well as environmental transformation underway in the basin. These transformations will likely accelerate as plans to alter regional river flows and flood regimes gather pace through increased infrastructural development, whether internally in Northeast Thailand (e.g. Nam Songkhram dam), transboundary (e.g. upstream dam building on the Mekong mainstream) or transbasin (e.g. Nam Ngum – Huay Luang water transfer project).

From the start, the E-flows study incorporated a wide range of actors within a single critical arena to discuss the potential linkages between flows, ecosystems and livelihoods and challenged many of the long-standing beliefs and notions that existed, often in official discourses used to justify particular large-scale infrastructure projects. This approach allowed a cross-pollination of ideas between local, regional and international experts from a variety of disciplines and institutional backgrounds, whilst opening up new modes of critical enquiry and thought. A senior provincial agricultural official claimed that the study had helped him understand that annual floods should be seen as beneficial to the ecosystem and communities, rather than as 'natural disasters' in need of an engineering solution, as claimed in dominant state narratives.

Through the field study component, the E-flows study made a contribution towards locally demonstrating the benefits of interdisciplinary approaches. The team of specialists did not only focus on their own disciplines, but actively engaged in jointly sharing knowledge, fieldwork tools and insights about the river floodplain ecosystem during the course of fieldwork.

The field study was novel in that it managed to time the field visits so that the team was able to witness and survey the river floodplain at the flood peak[12] and trough of the annual flood pulse phenomenon. The benefit of this timing for both visualisation and empirical grounding in relationships between flows, ecosystems and livelihoods is hard to underestimate and has theoretical and practical implications for any future studies that adopt a similar approach.

Finally, it can be stated on a modest level that the study has proven itself to be a useful, robust, economically efficient and participatory means to gaining insight into the socio-natural processes at play in river basin and water management in a complex milieu. As such, it is anticipated that the general approach and findings can provide useful lessons and pointers for any future efforts to apply E-flows approaches in other Lower Mekong River sub-basins. A weakness identified was the need for institutional backers and funders to keep behind the E-flows process and sustain the momentum beyond the initial period of implementation, perhaps by funding local partners to continue dialogues around the study findings. However, national politicians and senior decision makers in key water resources management bureaucracies such as the Department of Water Resources and Royal Irrigation Department will need to be convinced of the value of E-flows approaches, before there is any tangible change in Thai water allocation outcomes.

7.10   Conclusions

In the Mekong Region, there are various conceptions of E-flows in the research community and amongst policy makers and practitioners. E-flows implies different meaning to different people and disciplinary backgrounds. The authors found that in reviewing several cases in the Mekong Region, E-flows should not be seen as a one-size-fits-all approach, and this conceptual diversity is both a strength and a weakness. The cases presented in this chapter aimed to investigate the potential and limitations of E-flows and whether or not this approach can be used in informing negotiations about appropriate water regimes that also involves different stakeholders in the process. However, the question still remains as to whether implementing E-flows in the Mekong Region is an 'impossible dream' (IUCN 2005) or can be integrated into the basic approaches used by planners and decision makers tasked with managing river basins. There are clear differences between countries in the Mekong Region in terms of approach and understanding of E-flows. The E-flows pilot study in the Nam Songkhram River Basin in Thailand was applied at an intermediate level to ascertain appropriate water regimes in the river and identify

---

[12]It should be clarified that when water levels reach their wet season peak in the Lower Nam Songkhram River, the flows are not at their maximum, due to a backwater effect caused by the Mekong which may actually cause water flows to become quite slow or even stagnant for a few days, until water levels subside in the Mekong and flows resume in the Nam Songkhram.

stakeholder groups based on an interdisciplinary team of experts formed from academic, local government and community representatives. The case showed some of the challenges in linking E-flows theory and practice but placed a strong emphasis on stakeholder engagement, interdisciplinarity and dialogues as a contributing tool for piloting of the E-flows approach. In this way, the research differed considerably from the approach taken by the MRC's IBFM project, where subject specialists spent proportionately little time working alongside each other in the field or as a combined team. Though the IBFM project was designed to be multidisciplinary by having experts exchange knowledge regarding hydrological and social livelihood changes based on response to certain agreed indicators, the discussions were largely at a technical level (based on the specialist's field) not allowing for an interdisciplinary approach or a process of engagement with diverse stakeholders in the field.

In the case of Vietnam, a rapid E-flows assessment was carried out in the Huong River Basin amongst mainly government and international stakeholders, and the methodology for the assessment was quite different from that of the Nam Songkhram River Basin. Significant lessons from the Huong case were incorporated into the Nam Songkhram work, which was useful in experimenting further with the E-flows approach to determine its potential and limitations. Being a rapid assessment, the Huong River Basin case shows the limitations in enabling a truly interdisciplinary approach as was applied in the Nam Songkhram. Furthermore, the assessment had difficulty in linking the hydrological and ecological aspects. Whilst Vietnam has developed government policy by incorporating E-flows into policy at the national level, policy development has not been able to capture diverse perspectives from resources users in this basin. Some of the key government officials involved in the Huong River Basin study and the translation into Vietnamese can be seen as champions nationally of this approach, but more work is required to sustain these efforts.

The IBFM case employed by the MRC exemplified the need for a wide range of data to inform decisions on balancing economic and social benefits of development with environmental and social costs. Procedures developed for E-flows at the MRC focused on 'minimum flows' but lacked a process agreed by relevant stakeholders to use the results of the IBFM in a more holistic manner that could have provided a framework for negotiating water-related decisions.

These cases exemplify that there are both opportunities and challenges to employing the E-flows approach in the Mekong Region. There are inherent problems of unequal power distributions exemplified in the case examples along with inherent tensions between environmental, social and economic agendas. This is not only as an abstract concept in legislation that may be vulnerable to different interpretations and thereby impossible to become effective in law enforcement, but as a process that involves multiple stakeholders dialoguing to determine the best possible flow regime for the Mekong Region. As the multi-stakeholder translation process of the book *Flow* showed, capacity and constituency building and, equally importantly, finding the correct translated terms for complex scientific concepts are crucial elements to enabling people to participate effectively in an E-flows process.

In short, for E-flows processes to be used in the Mekong Region, there needs to be a sustained process of support and trust building between numerous interested actors and institutions to build a critical mass of expertise and understanding during a sustained period of conceptual internalisation and practical demonstration.

Acknowledgements This research paper is from M-POWER's Improving Mekong Water Allocation Project (PN67) (see: www.mpowernetwork.org), part of the CGIAR Challenge Program on Water and Food (CPWF). The work was made possible by a financial contribution to the CPWF by the European Commission, overseen by the International Fund for Agricultural Development (IFAD). We would like to acknowledge the teams of people that participated in the E-flows experiments discussed in the paper, with Jessica Illaszewicz at IUCN Vietnam being deserving of a special mention. We would also thank the many people who have provided support to produce this paper including Megan Knight, who initially drew together research on this topic which provided a basis for this paper, Marko Keskinen, Jeffrey Richey, Chu Thai Hoanh and Francois Molle for reviewing earlier drafts, Matti Kummu for kindly generating a high-quality map and Rajesh Daniel for editing and compiling all of the documentation for PN67. The paper is dedicated to one of the authors, David Hall, an excellent colleague and friend, who was sadly taken away from us by an accident in Lesotho in late 2010.

References

Arthington AH, Zalucki MJ (1998) Comparative evaluation of environmental flow assessment techniques: review of methods. Land and Water Resources Research and Development Corporation, Canberra

Baran E, Myschowoda C (2009) Dams and fisheries in the Mekong basin. Aquat Ecosyst Heal Manag 12(3):227–234

Blake DJH, Friend R, Promphakping B (2009) Landscape transformations and new approaches to wetlands management in the Nam Songkhram river basin in northeast Thailand. In: Molle F, Foran T, Käkönen M (eds) Contested waterscapes in the Mekong region: hydropower, livelihoods and governance. Earthscan, London, pp 173–202

Brisbane River Symposium (2007) Brisbane declaration. www.riversymposium.org. Accessed 15 July 2010

Dore J (2010) Multi-stakeholder platforms. In: Dore J, Robinson J, Smith M (eds) NEGOTIATE: reaching agreements over water. IUCN, Gland, pp 37–58

Dore J, Robinson J, Smith M (eds) (2010) NEGOTIATE: reaching agreements over water. IUCN, Gland

Dyson M, Bergkamp G, Scanlon J (eds) (2003) Flow: the essentials of environmental flows. IUCN, Gland

Environmental Flows Network (2004) Environmental flows newsletter: environmental perspective on river basin management in Asia. 1(1) www.eflownet.org

Environmental Flows Network (2006) Environmental flows newsletter: environmental perspective on river basin management. 3(1)

Friend R, Blake DJH (2009) Negotiating trade-offs in water resources development in the Mekong basin: implications for fisheries and fishery-based livelihoods. Water Policy 11(Suppl 1):13–30

Global Water Partnership (GWP) (2000) Integrated Water Resource Development. GWP Technical Advisory Committee Background Paper 4. GWP, Stockholm, 71 pp

Guttman H (2006) River flows and development in the Mekong river basin. Mekong Update and Dialogue, Australia Mekong Resource Centre. 9:3 July - December, Sydney

Hirji R, Davis R (2009) Environmental flows in water resources policies, plans and projects: findings and recommendations. The International Bank for Reconstruction and Development and The World Bank, Washington

IUCN (2005) Environmental flows – ecosystems and livelihoods – the impossible dream? Session report. 2nd Southeast Asia Water Forum. IUCN Water and Nature Initiative and the Mekong Wetlands Biodiversity Conservation and Sustainable Use Programme, Bali, Indonesia, p 39, 29 Aug– 2 Sept 2005

IUCN Vietnam (2005) Environmental flows: rapid environmental flow assessment for the Huong River Basin. IUCN Vietnam, Central Vietnam, p 82

IUCN, TEI, IWMI, M-POWER (2007) Exploring water futures together: Mekong region waters dialogue. Report from regional dialogue. The World Conservation Union, Thailand Environment Institute, International Water Management Institute, M-POWER water governance network, Vientiane, Lao PDR, p 75

International Water Management Institute (IWMI) (2005) Environmental flows: planning for environmental water allocation. Colombo, Sri Lanka: International Water Management Institute (IWMI). 6 p. (IWMI Water Policy Briefing 015)

Junk WJ, Wantzen KM (2004) The flood pulse concept: New aspects, approaches and applications – an update. In: Welcomme RL and Petr T (eds) Proceedings of the 2nd international symposium on the management of large rivers for fisheries, vol 2. Food and Agriculture Organization, Bangkok, pp 117–140

King J, Brown C (2009) Integrated basin flow assessments: concepts and method development in Africa and southeast Asia. In: Freshwater biology. Blackwell Publishing, Oxford

King J, Brown C, Sabet H (2003) A scenario-based holistic approach to environmental flow assessments for rivers. River Res Appl 19:619–640

Klein JT, Newell WH (1997) Advancing interdisciplinary studies. In: Gaff J, Ratcliff J (eds) Handbook of the undergraduate curriculum. Jossey-Bass, San Francisco, pp 393–394

Krchnak K (2006) "Greening" hydropower: integrating environmental flows considerations. The Nature Conservancy, Arlington

Kummu M (2010) Map of pilot basins with E-flows experimentation in the Mekong Region

Ministry of Natural Resources and Environment (2006) National water resources strategy towards the year 2020. Culture-Information Publishing House, Hanoi

Molle F, Floch P, Promphakping B, Blake DJH (2009) "Greening isaan": politics, ideology, and irrigation development in northeast Thailand. In: Molle F, Foran T, Käkönen M (eds) Contested waterscapes in the Mekong region: hydropower, livelihoods and governance. Earthscan, Londong, pp 253–282

Moore M (2004) Perceptions and interpretations of environmental flows and implications for future water management – a survey study. Master thesis, Department of water and environment studies, Linkoping University, Sweden

MRC (Mekong River Commission) (2005) Overview of the hydrology of the Mekong basin. Mekong River Commission, Vientiane

MRC (Mekong River Commission) (2006a) Procedures for the maintenance of flows on the mainstream. Mekong River Commission, Vientiane

MRC (Mekong River Commission) (2006a) Integrated basin flow management report no. 8. Flow-regime assessment. Mekong River Commission, Vientiane

MRC (Mekong River Commission) (2008) Integrated basin flow management progress report: social assessment team June – August 2007. Mekong River Commission, Vientiane

Osborne M (2007) The water politics of China and Southeast Asia II: rivers, dams, cargo boats and the environment. Report to the Lowy Institute. http://www.lowyinstitute.org/Publication.asp?pid=589

Richter B, Warner A, Meyer J, Lutz K (2006) A collaborative and adaptive process for developing environmental flow recommendations. River Res Appl 22:297–318

Tharme RE (1996) Review of international methodologies for the quantification of the instream flow requirements of rivers. Department of Water Affairs and Forestry, Pretoria

Tharme RE (2003) A global perspective on environmental flow assessment: emerging trends in the development and application of environmental flow methodologies for rivers. River Res Appl 19:397–442

Tharme RE, Smakhtin VU (2003) Environmental Flow Assessment in Asia: Capitalizing on existing momentum. In: Proceedings of the First Southeast Asia Water Forum, November 17–21, 2003, Chiang Mai, Thailand. Volume 2: 301–313. Bangkok, Thailand: Thailand Water Resources Association

Vidal A, van Koppen B, Blake D (2010) The green-to-blue water continuum: an approach to improving agricultural systems' resilience to water scarcity. In: Lundquivst J (ed) On the water front: selections from the 2009 world water week. Stockholm International Water Institute, Stockholm

WWF (2009) The greater Mekong and climate change: biodiversity, ecosystem services and development at risk. WWF Greater Mekong Program, Thailand

Chapter 8
IWRM as a Participatory Governance Framework for the Mekong River Basin?

Philip Hirsch

Abstract Integrated water resources management (IWRM) has been widely adopted as an over-arching framework for managing river basins. However, tensions are inherent in IWRM between top-down and bottom-up approaches to management. In seeking to move away from fragmented toward more integrative approaches to bio-regional natural resource management at the level of the river basin, IWRM initiatives also tend to centralise. Yet a participatory ideal, or at least rhetoric, is seen in "stakeholder-based" and other more inclusive approaches to basin management. In the Mekong, these approaches are seen in the Mekong River Commission's basin development stakeholder processes, subarea-based planning, and the establishment of river basin organisations. These are essentially top-down driven approaches to participation. On the other hand, some regional NGO initiatives, broad coalitions such as "Save the Mekong", community-based networks such as the 3SPN network in Cambodia, and decentralised irrigation management in its various forms, apply participation from the ground up and often seek to challenge projects that result from immense pressures for development of the river and its tributaries for hydropower. We need to move our understanding of IWRM in river basin governance away from a technical, "best practice" approach, toward recognition of its inherently political nature and its embeddedness in cultural practices at various levels.

8.1 Introduction

On 1 April 1996, the International Rivers Network wrote to the Mekong River Commission's first Chief Executive Officer, Yasunobu Matoba, to provide feedback on a recent meeting to discuss the new organisation's approach to public participation. An excerpt from the letter is instructive:

P. Hirsch (✉)
School of Geosciences, The University of Sydney, Madsen Building (F09), Rm 472,
Sydney, NSW 2006, Australia
e-mail: philip.hirsch@sydney.edu.au

J. Öjendal et al. (eds.), *Politics and Development in a Transboundary Watershed:*
The Case of the Lower Mekong Basin, DOI 10.1007/978-94-007-0476-3_8,
© Springer Science+Business Media B.V. 2012

It was most discouraging to learn... that the MRC has absolutely no plans to incorporate public input into its activities. You stated, for example that participation 'is not the responsibility of MRC, but of the member countries'... You asked rhetorically, "How are we supposed to do it? We can't allow everyone to come to our meetings" (Open letter signed by Owen Lammers and Rani Derasary).

Organisations with broad-scale governance roles often find themselves targeted critically for their top-down approach to management and lack of attention to local participation. In a transboundary river basin context, critiques of macro-oriented governance perspectives and practices target organisations such as the Mekong River Commission (MRC). These critiques raise the question of how one "does" participation in such a basin. The question has been raised with increasing frequency as MRC has established various programs that have sought either to engage with "stakeholders" or to establish analyses, structures and approaches that take its work to a more local level. Tensions are evident between MRC's mandate for transboundary management, under which everything within individual countries is seen as outside its ambit and risks impinging on national sovereignty, and the "participatory turn" that comes with the package, adopted by MRC and many other river basin organisations, referred to as Integrated Water Resources Management (IWRM).

The tension between top-down and bottom up approaches is inherent in IWRM at a conceptual as well as at a technical or political level. IWRM is by definition and terminology integrative. On the one hand, being integrative means seeing and acting on the whole picture in a geographical sense (entire basin as an integrated system), across sectors, and in pursuit of multiple goals in a holistic way so that economy, environment and society are treated as interdependent rather than separate or competing considerations in the use, management and conservation of water resources. In this sense, IWRM has totalising tendencies and invites big-picture thinking in place of locality-based attention. On the other hand, integration means incorporating diverse stakeholders from different sectors, localities, countries, interest groups and so on. This more inclusive dimension of IWRM tends toward participation, attention to the integrated nature of livelihood and resource systems at multiple scales including the localities where most people in a river basin such as the Mekong live out much of their lives, and negotiation, accommodation or management of different views, interests and value systems. There are thus both centripetal and centrifugal tendencies in IWRM.

Few IWRM-based programs see the tension as an insurmountable problem, since participation can be practiced in so many different ways and means so many different things to different people and agencies. This chapter delves further into participation and IWRM from above and below. It commences by revisiting the problem area – fragmentation – out of which IWRM emerged, asking what this means for participation. The discussion then goes on to examine ways in which MRC has responded to critiques of a non-inclusive and overly centralised approach in its BDP and other IWRM-related program areas. The mirror of this is the set of NGO and "grassroots" or "community-based" actions that have upscaled their concerns about local livelihoods to a basin-level frame of reference, and several of these initiatives are also examined. The chapter concludes with a critical discussion of the rendering of participation, and indeed of IWRM, to the technical level in a field that is inherently political and embedded in cultural practices, and of the associated tendency to adopt birds' eye perspectives at the expense of a more emic appreciation of everyday concerns of the basin population.

8.2   Participation, FWRM and IWRM

Just as participation is an implicit critique of autocratic, centralised, non-consultative, hierarchical decision making and governance, so Integrated Water Resource Management is implicitly a critique of management approaches that treat different parts of a system as if others did not exist. We can refer to the object of critique as Fragmented Water Resources Management (FWRM). Historically, FWRM has taken a number of forms. A common form of fragmentation is the division of the administration of natural resources into different line agencies, who fail to talk to one another, who compete for budgets, and who often demarcate their sphere of responsibility territorially, in ignorance or denial of the inter-related nature of resource and livelihood systems in any one locality, and in the case of water the land-vegetation-water interactions that are part of the functioning of natural systems. This is sometimes referred to as the silo effect. Another form of fragmentation is the geographical division of natural resource systems, sometimes referred to as bio-regions, along administrative and political boundaries so that there is lack of coordination, and sometimes competition or conflict, across these boundaries. Transboundary river basins are a case in point. Other fragmentary approaches to water resources management include the gap between high level decision makers and "trustees" (Li 2007) of natural resource governance, on the one hand, and the local resource users whose livelihoods and everyday existence is directly dependent on water and other resources, on the other. There is further fragmentation when water and other resources are managed by some for economic growth purposes, narrowly defined; by others for conservation without reference to changing needs and aspirations of resource users; and yet others for fulfilment of social goals without reference to the physical limits of the resource system. In all these senses, IWRM offers a seemingly logical and promising umbrella under which a more inclusive, "triple bottom line" approach can enhance social equity, environmental sustainability and economic efficiency in the way water can be managed. IWRM is primarily considered in a river basin context, since large and small catchments define the boundaries of integrated natural systems within which water is collected, flows and interacts with land and vegetation.

Participation has seen a discursive and political slide from being a critical, even subversive challenge, toward conventional practice in a mainstreamed orthodoxy. The multiple interpretations of participation range from the more substantial and radical to the more cynical and vacuous (Arnstein 1971). Cooke and Kothari (2001) have shown how this ultimately leads to "new tyrannies" of power in the name of participation. Similarly, IWRM has seen a slide from the margins when Total

Catchment Management and bio-regionalism were considered critical challenges to the administrative boundaries that paid no attention to natural resource systems and bureaucratic silos within which natural resources are governed in isolation from one another, in ignorance of ecosystem connectivity and interactions. Instead, IWRM risks becoming a multi-criteria checklist notion (Biswas 2004), a "nirvana concept" (Molle 2008) or simply a feel-good mantra or window dressing for programs and institutions conducting business as usual.

If both participation and IWRM have seen a slide from critique to mainstreamed and anodyne discursive covers for business as usual, how are we to understand the role of participation as a progressive ideal within an IWRM approach that is both basin-wide and seeks to incorporate the concerns, needs, values, livelihood realities and aspirations of those at the grassroots level? The case areas that follow investigate a number of attempts to "downscale" from the top down, or "upscale" from the bottom up (Sheldon 2005).

8.3   Governing Through Participation from Above …

MRC is normally considered a governed rather than governing organisation. It is owned by the riparian member governments, who govern through the MRC Council at Ministerial level and Joint Committee at the level of Permanent Secretary of the ministries nominated by each country to manage Mekong Basin related affairs. MRC is not a regulatory agency with statutory authority to force its members to conform to particular courses of action. Rather, it is the implementing agency for the 1995 *Agreement on the cooperation for the sustainable development of the Mekong River Basin*, a loose set of principles based on Articles that are subject to interpretation and some of which have been elaborated into more detailed Procedures. Negotiation of the Agreement and development of rules and procedures has paid little or no attention to involvement beyond the "expert" level within each riparian country, and has in fact relied considerably on international consultants. MRC's role as a knowledge-based support agency, as a planning agency, as a forum for discussion by governmental or wider stakeholders, or as an investment facilitator remains ambiguous, but there is general agreement within and beyond the organisation that MRC has a relatively weak role in governing water-related decisions within the Mekong River Basin. In part this is because national interests as articulated by member governments take precedence over basin-wide considerations. In part it is because, without China as a member, MRC is not relevant to the upper half of the river's length; in part it is because of the consensus-style politics that govern MRC within the wider ASEAN transnational political culture; and in part it is due to the lack of embeddedness of MRC within the governance and societal affairs of its constituent member countries (Hirsch et al. 2006).

Despite – and in some cases because of – MRC's weak position vis-à-vis its individual constitutive member states, the organisation has been seen as aloof, removed, out of touch with the needs and perspectives of farmers, fishers and wider civil society within the lower Mekong River Basin (ibid). In part, this has been the product of styles of leadership (Hirsch 2008), and in part it has been due to the combination of governance structure and donor dependence and domination by international consultants that orient MRC more toward foreign development assistance organisations and toward those Ministries represented on the Council and Joint Committee than toward the millions of basin resource users – the fishers and farmers who make up the bulk of the lower Basin's 70 million people.

However, in recent years, MRC has seen a slow but perceptible participatory turn in its style and in the substance of some of its programs. MRC's website now mounts many more documents, in a more accessible and timely fashion, than previously. In part this is no doubt once again a result of donor pressures, for example as stipulated in an Organisational Review in 2006 (MRC 2007) that recommended greater accessibility of the Commission to people and organisations living in the basin, better participatory processes and structures, and above all a "riparianisation" of the Commission. In part it is the product of a more open regime with the change of Chief Executive Officer in 2008. However, the greater open-ness is also a response in part to the demands of an increasingly vocal and well-informed civil society that the organisation should broaden it approach and reach out to a wider range of interest groups in the four lower Mekong countries. Three areas of MRC's work illustrate this turn, and also indicate the limits to its participatory substance.

8.3.1   Meeting-Based Stakeholder Engagement: Basin Development Plan Forums

One way in which MRC has appeared to engage in a more participatory way than hitherto is by opening up its key processes to a wider range of feedback from different groups. The Basin Development Plan (BDP) Stakeholder Forums represent the most extensive form of "stakeholder engagements". The Forums, held in Vientiane in March 2008, in Chiang Rai in October 2009 and in Vientiane in July 2010 each invited a range of academic, government, NGO, international agencies, donors, consultants and others deemed to have a "stake" in the BDP outcomes to meet over 2 days and discuss the direction of the BDP2 program. The ostensible aim was to inform the wider basin community of developments in the basin development planning, and to secure feedback to improve the planning process.

The BDP Forums were a departure from a previously quite closed project-based planning process. They put a large number of documents on the table, and opened the BDP planning process to more critical feedback than hitherto. While the majority of the presentations at the third and final BDP Stakeholder Forum came from the BDP team itself or from other sections of the MRC Secretariat (MRCS), the meeting also invited three critical reviews. One was an analysis of the BDP hydrological assessment, carried out by a member of the Integrated Knowledge Development Program of MRCS; the second was a review of the BDP environmental assessment, carried out by the WWF Greater Mekong Program;

the third was a critique of the IWRM-Based Basin Development Strategy for the Lower Mekong Basin, carried out by the author of this chapter in his role as an independent academic (Hirsch 2010). Question and answer sessions allowed for some critical feedback.

Despite the innovation of stakeholder forums, there are significant limitations in terms of enhancing participatory governance. MRCS maintains control over who is deemed a significant stakeholder, who is invited to attend, who is invited to speak, and whose commentary is published in final proceedings. An interesting case in point at the third forum is that World Fish were initially invited to present on the fisheries implications of the development scenarios being put forward through the BDP program. However, after submitting a hard-hitting analysis, based on the work done on fisheries impacts of mainstream hydropower dams for the MRC Strategic Environmental Assessment (SEA), World Fish was told a few days ahead of the meeting that there would be no room to present on that topic, since the SEA was already being presented by MRCS' hydropower program. World Fish was instead invited to talk about community fisheries management. In the event, the MRCS presentation on the SEA was about the process and not the findings, in effect – and perhaps by intention – censoring out the most critical scientific findings relevant to the substance of the consultation. A further limitation lies in the nature of interaction at a forum that presents highly condensed technical information with no chance for participants to see and digest it ahead of time. My own critique was of a draft strategy document from 9 months previously, an updated version of which was presented to the participants – and myself – on the day of the meeting after having shared my critique with the organisers a few days prior. This managed interaction gives "stakeholders" a sense of pre-ordained agendas, despite the relatively relaxed space for critical interventions in the meeting hall itself. This key limitation to stakeholder-based natural resource management reflects wider concerns over manipulation and marginalisation that arises when playing fields are only temporarily levelled for the period of consultation (Edmunds and Wollenberg 2001). The final summary given by the Director of the BDP at the end of the conference gave a false sense of consensus, as if the meeting had given the program in imprimatur to move forward on its preferred scenario and strategy. This tendency as a danger to meaningful stakeholder participation was clearly identified in MRC's own report into stakeholder participation:

> If the stakeholder consultation event moves too quickly to the final discussion (of specific development strategies) the space for meaningful debate is already closed, and it is difficult for participants to critique the background analysis, or to raise alternative development strategies (MRC 2010a: 19).

Other than the limitations in consultation processes, the substance of the IWRM-Based Basin Development Strategy document itself is instructive regarding the way in which IWRM is conceptualised in Basin planning vis-à-vis participation. The 68 page draft document tabled at the July 2010 Forum refers to IWRM no less than 137 times. Other than reference to taking an "integrated basin perspective", the document does not define IWRM. However, it refers, in passing, to

MRC's triple bottom line objective of achieving "an economically prosperous, socially just and environmentally sound Mekong Basin", i.e. integrating economic, social and environmental considerations, and to providing a framework into which project developers (mainly dam proponents and financiers) can place their proposals with a degree of certainty. There is little in the document that sets a strategic view of participatory input into decision making. Rather, the IWRM-Based Basin Development Strategy is an integration of data from a range of studies carried out by MRCS that seeks consensus on acceptable scenario. The draft Strategy pushes a scenario that includes 6 of the 11 lower mainstream dams, in addition to 56 tributary dams (26 already under construction or approved, and another 30 that are planned) and to the existing or planned eight mainstream dams on the Lancang Jiang section of the river in China. Any "consensus", therefore, would be limited to the national representatives who form the Joint Committee and Council of MRC, and would certainly not include the wider set of voices – and the weight of scientific opinion – who remain steadfastly opposed to any mainstream hydropower development.

8.3.2   Nested Hierarchies: Sub-area Planning as a Surrogate for Everyday Governance

MRC has recognised its degree of removal from the myriad locales where most people in the Basin live out their daily lives. Through the BDP, this issue has been dealt with in a nested hierarchical fashion, by dividing up the basin into a number of sub-areas. The demarcation of these sub-areas comprises a combination of natural boundaries of sub-basins and political boundaries, such that each sub-area is within a single national space, and sub-areas are based on groups of tributary basins that are truncated at the national borders. Each sub-area is treated as a unit for planning at the sub-basin level. The key aim of the sub-area analysis is to identify priorities around water and related natural resource management within each of these areas. Forums are organised to bring together key stakeholders to work with consultants assigned through the BDP program in coming up with sub-area analysis reports. In some cases, these reports identify desired water resource developments. In others, they articulate concerns over actual or potential impacts of water and related land-based developments.

The composition of sub-area committees is heavily government-based, and their administration is consultant-managed rather than locally generated. For example, out of 82 named participants in a sub-area meeting for the 7L (Sekong) sub-area in Laos, held in Champassak in February 2010, only two were village-based. Four participants were from the consultants' team running the meeting, five from MRCS, five from NGOs, two from international agencies, one from the media, and 63 from National, Provincial and District government (MRC 2010b). One of the recommendations of MRC's stakeholder analysis is that the sub-area planning should be more inclusive of civil society and ordinary villagers' voices (MRC 2010a).

While the question of representation is important, particularly where there is a tendency to limit "stakeholders" to governmental and sometimes NGO agencies rather than include farmers and fishers in the sub-area planning process, a more fundamental question is the connection between sub-area planning and the everyday governance that influences decisions relevant to water resource development and management. As a project-based activity, BDP tends to have little purchase in the regulatory or other governance arrangements relevant to key decisions. The proliferation of tributary dams in Lao PDR with little or no reference to 10 years of basin development planning is evidence of the disconnected processes of sub-area planning, on the one hand, and the fundamental decisions being taken on the future of most of the Mekong's sub-basins on the other.

8.3.3   Bioregional Subsidiarity: RBOs as the Answer, but What Was the Question?

Closely related to the sub-area approach is a more generic mode of governance under IWRM principles at the sub-basin level, which is the establishment of river basin organisations (RBOs). RBOs, also referred to as River Basin Committees or Catchment Management Agencies, are the institutionalised mirror of MRC at a sub-basin level. The RBO concept has its origins in the critique of territorial management that follows administrative rather than natural boundaries, and as such can be seen as a kind of bio-regionalism. Bio-regionalism in the Mekong Region has a history that is more commonly associated with radical NGO critique of territorially fragmented management than with mainstream territorial management (Thongdeelert 1997). In principle, RBOs bring together a range of stakeholders from within the catchment for activities including sub-basin planning, water allocation, securing of project funding for water and related resource development projects, and in some cases for liaison with neighbouring basins and with higher level basin authorities.

The principal agency promoting RBOs in the Mekong region and more widely throughout Asia is the Asian Development Bank, which together with the Japan Water Agency and Tokyo-based Asian Development Bank Institute has also established a Network of Asian River Basin Organisations (NARBO). NARBO states its goal as being "… to help achieve IWRM in river basins throughout Asia"[1]. ADB has also supported the drafting of national legislation in each Mekong country that provides for RBOs and which more recently has been adopted by relevant bureaucratic agencies and regularised in all four lower Mekong countries. In Thailand, 29 River Basin Committees were established by the Department of Water Resources in 25 river basin areas carved out of the national territory. One of the agendas for IWRM employing these basins and RBCs as the framework for governing them has been the inter-basin transfer of water from "water surplus" to "water deficit" basins, a highly

---

[1] www.narbo.jp

controversial river-linking program under the generic title of "Water Grid" (Molle and Floch 2007). In Laos, RBOs are less generic and have been established in just a few river basins, with the most active example being the Nam Ngum River Basin Organisation, a project funded with ADB assistance. Similarly, in Vietnam only a few RBOs have been established, including the Red River, Dong Nai and Srepok Basins, all of which are significant for hydropower development. In Cambodia, the only Basin organisation established to date is the Tonle Sap Authority.

The RBOs with functioning secretariats in the Mekong Region are mostly funded by bilateral or multilateral development agencies. The Srepok River Basin Organisation, for example, was funded by Danida, the Danish development agency, in Vietnam. Even though the catchment extends into Cambodia, and the major upstream-downstream issues in the Srepok Basin are the impacts of Vietnam's hydropower development and groundwater extraction for coffee cultivation on the downstream indigenous communities in Cambodia, the RBO is truncated at the national border. The Tonle Sap Authority is a high level organisation under the Minister for Water Resources, with authority in the eight provinces surrounding Tonle Sap Lake, but with a primary area of focus on the area inundated by the lake during the wet season, bounded by National Roads 5 and 6, running south and north of the Lake respectively. The authority has no participatory structure, but is rather based on management through zoning of conservation, fishing and agricultural areas.

In summary, the experience of river basin organisations as a stakeholder-oriented devolved governance arrangement in the Mekong River Basin is that they are largely consultant-designed and driven, and have been initiated through development assistance project funding rather than out of problem-definition by the key stakeholders in each basin. A series of presentations on river basin organisations at the third Stakeholder Forum of MRC's Basin Development Plan program described their structures, funding base and roles in terms of strategic planning, conflict resolution, coordination of development and natural resource management activities within their zones of jurisdiction, but none had a concrete issue area or local involvement that indicated a problem to which they were a potential solution.

8.4 … And Governance Through Participation from Below

While there has been increasing lip-service, if not effective practice, on the part of mainstream agencies in the field of participation, stakeholder involvement and devolved management to promote IWRM as a more inclusive and less fragmented approach to basin management, civil society organisations have promoted and practiced a different kind of participation in basin-wide issues. From the point of view of the more critical NGOs with Mekong-wide purviews, the point of civil society participation is not limited to management by local communities within their own vicinity. On the contrary: the integrated nature of the basin, and the recognition that impacts in one part of the basin (in particular upstream) have effects over long distances (especially downstream) informs an advocacy approach in which networked community-based and issue-based groups have come together in a number of campaigns.

Some of the "participation from below" takes the form of issue-specific actions and agendas. Other initiatives look toward a more established civil society governance framework, mostly along a network model. Few of these initiatives frame themselves in terms of "participation" per se, and even fewer employ the IWRM discourse. Yet in substantive terms, they can be seen as integral to a stakeholder-oriented approach to integrated river basin management.

8.4.1 Basin-Wide Representation: Regional NGOs as People's Organisations?

Civil society voices in the Mekong have grown mainly out of Thai-based experience in mobilising against destructive large scale projects. The Thai environmental movement's success during the late 1980s in halting the construction of the Nam Choan Dam and in having logging concessions cancelled had two main effects (Hirsch and Lohmann 1989). One was to build confidence and awareness within Thailand, seeing a proliferation of environment-based activism, albeit one that took Thailand's environmental movement in a number of different directions (Hirsch 1997). The other was to shift Thailand's resource-based investment across national borders into neighbouring Mekong countries, where resources remained abundant and where there was little space for endogenous environmentalism to grow. Two Thai-based organisations followed this move by extending their area of concern to the wider Mekong Region: Toward Ecological Recovery and Regional Alliance (TERRA), and the Southeast Asia Rivers Network (SEARIN).

Both TERRA and SEARIN have targeted MRC, ADB, the World Bank and international donors for their support of hydropower development in the Mekong. The main strategy has been to build a network of regional partners in Vietnam, Laos and Cambodia, a challenging task in countries without the political space for NGO activity that has long been part of the Thai NGOs' working environment. Networks have included projects supported by international NGOs, universities, and in some instances individuals in local and national government agencies. While the concerns of these "regional civil society" networks are multifold, there has been a heavy emphasis on opposition to construction of large dams. Some high profile campaigns include the ultimately unsuccessful attempt to stop the World Bank from supporting construction of the Nam Theun 2 dam, and the exposure of grievances of those affected by the ADB-funded Theun Hinboun Dam. Activism has taken many forms, including a series of conferences and dialogues that have mobilised public opinion, with the assistance of the media, through awareness raising and drawing on Thailand's experience as a warning. A number of sign-on letters have been directed at MRC and ADB, directly and by attempts to influence the major donors – and the tax-paying public of the countries from which they draw their development assistance funds – to withdraw program funding or set conditions.

One of the challenges facing this sort of civil society activity is the question of representation. While a number of sign-on letters have started, "We the people of the Mekong…", the organisational basis of TERRA, SEARIN and its albeit wide network of regional and international partners is not representative in a strict governance sense. Certainly the locally based projects, organisations and grievance-derived networks (see-below) that provide the breadth and local engagement capacity of the network are closer to the "ground" of village-level opinion and concerns – at least in some localities – than IUCN and WWF, the international "civil society" organisations that have observer status on the Mekong Council. But there remains no systematic framework for representation of community interests to match, for example, the Murray Darling Basin Authority's Basin Community Committee (formerly known as Community Advisory Committee to the Murray Darling Basin Commission).

8.4.2   Issue-Oriented Advocacy: Save the Mekong

While there is no permanent or institutionalised basin-wide representative input into the IWRM-based planning for the Mekong, there has been a high degree of mobilisation around key issues. One of the most galvanising issues has been the revived move to build dams on the lower Mekong mainstream. For many years, these dams were assumed by NGOs and by many of the international donors to MRC to be off the agenda, since they are simply too damaging, especially to the migratory fish on which the Mekong population depends for between 40% and 80% of their animal protein intake, and which provides the basis not only for the larges freshwater fishery in the world, but also the second most biodiverse river system on earth. The threat to this fishery has long been asserted by NGOs, and confirmed by studies conducted by MRC's own fisheries program. The projected impact of the mainstream dams has been quantified by the Strategic Environmental Assessment commissioned by MRC as a key plank of its IWRM strategy.

In 2008, a group of NGOs and civil society activists banded together to form the Save the Mekong (StM), bringing together an exceptionally wide group of NGOs, artists, academics, people from dam-affected or threatened communities, and others. In a move to show widespread opposition to any dams on the mainstream, other than those which are a fait accomplit on the Lancang Jiang in China, StM ran a postcard campaign that delivered more than 23,000 signatures in a letter to the Prime Ministers of Cambodia, Lao PDR, Thailand and Vietnam. The campaign included a travelling photographic exhibition with associate public lectures and seminars in the four Mekong countries, in Europe and Australia, targeting countries whose governments support MRC through their aid programs. The campaign also attracts widespread media coverage, keeping the issue alive in the wider public consciousness. It receives support from a number of international NGOs, including Oxfam Australia, International Rivers, and Mekong Watch (Japan).

While StM has managed to garner wider and more sustained public support than most Mekong-oriented advocacy initiatives, it has done so by focusing on a single issue – probably the one that has greater significance for the Mekong's future than any other. Despite their widespread impacts, the Mekong Mainstream Dams cannot be said to be the basis for an IWRM framework per se. As such, therefore, it appears that the most effective mobilisations "from below" are not necessarily those that work within an IWRM context, and certainly they do not employ or dwell on IWRM discourses. On the contrary, it is the singular focus of the campaign that gives it effect.

8.4.3   River Basin Networks: 3SPN

Below the scale of the entire river basin, participation from below at the sub-basin level has been mobilised around problem-driven issues. The most substantial of these is the response to impacts of the Yali Falls Dam in Vietnam during the late 1990s, whose sudden water releases and closures had major transboundary impacts on mainly indigenous minority communities in the northeastern Cambodian provinces of Ratanakiri and Stung Treng. The impacts and the response to them have been quite extensively documented elsewhere (Baird 2002; Hirsch and Wyatt 2004). The significance of the case for the current discussion is the mode of organisation of river basin communities into a network that has meaning for those dependent on the Se San River.

As the impacts of Yali Falls Dam became apparent, it was initially external NGOs who brought to the attention of isolated river-bank communities that the unexplained and unprecedented fluctuations in river levels, which had led to loss of life, livestock, river bank gardens and property, were due to human-induced and not natural events. The non-timber forest products project in Ratanakiri used its concern for livelihoods to assist in the establishment of a community network under the flag of the Se San Protection committee. This has since expanded to incorporate the two neighbouring transboundary basins whose rivers flow westward from Vietnam (in the case of Srepok) and Laos (in the case of Sekong) to a confluence with the Se San before the system drains into the Mekong at Stung Treng town. The 3S basin, as it has become known, is the basis for the 3S Protection Network, or 3SPN, and it has also been emulated as a geographic entity by an ADB program for IWRM that in principle will support a more stakeholder-oriented approach to planning for the basin. However, while the consultants have been doing their planning studies, the upper Se San and Sre Pok have been developed so rapidly that most of the economically attractive hydropower dams (about six on each tributary) have now been completed or are under construction.

Early on, 3SPN and its predecessor network attempted to assert upward negotiating pressure on MRC to intervene, on the basis that MRC is a transboundary river basin agency. However, MRC insisted that any intervention must be based on a Cambodian government request, since the Commission's members are its governments and not the people whom they govern. MRC did set up a bi-lateral committee to investigate the Se San problem, but this went into abeyance after three meetings and no compensatory or mitigatory action on the part of the owner and manager of Yali Falls Dam, Electricity of Vietnam. EVN acknowledged the unintended impacts and some progress has been made in improving information on dam releases upstream. However, permanent loss of fisheries and water quality in the Se San river mean that ultimately this instance of participation from below has yielded rather little by way of concrete benefits for those involved.

8.4.4 Everyday Practice: Irrigation Associations

A final instance of participation from below is the range of practices of irrigators in the Mekong Basin. From about 2000, Laos and Cambodia have been embarked on a rapid increase in irrigated agriculture within their respective parts of the Basin, and Thailand and Vietnam have longer standing irrigation developments within the Basin and elsewhere in their respective national territories. Thailand, Vietnam and Laos have a long tradition of irrigated agriculture in mountainous valleys, typically relying on weir and channel (*muang-faai*) arrangements that are in turn based on a specific set of culturally embedded governance arrangements for their construction, maintenance and water allocation.

One way of looking at *muang-faai* arrangements is to see them as the original IWRM, albeit writ small. Rather like the Subaks that Geertz and Lansing described in Bali in terms of their holistic ritual significance (Geertz 1972; Lansing 1987), many such schemes constituted far more than water management organisations and associated infrastructure. They included reference to headwater forests, systems of belief in spirits at the head of the stream from which the channels took water, and systems of authority, respect and representation that formed the basis for village social organisation. Later, NGOs would take *muang-faai* as a symbolic model of local autonomy and capacity for integrated natural resource management (Thongdeelert 1997), despite evidence that the localism implied in this somewhat romanticised view was quite a problematic notion (Cohen and Pearson 1998).

Elsewhere, participatory irrigation management and development has been fostered in a more programmatic way. Indeed, the acronym PIMD is an indicator of the mainstreaming of local irrigation management as a part of IWRM-oriented practice. In Cambodia, for example, Farmer Water User Communities have been set up to manage the rapid expansion of irrigation, either on new sites or through the rehabilitation of Khmer Rouge era projects. However, in the absence of experience with collective management of water, collection of the requisite irrigation service fees, and unconnected as they are from existing authority structures, the FWUCs, as they are termed, have often come to grief along with the infrastructure they are supposed to manage (Chou 2010; Daravy 2010).

8.5 Conclusions: Beyond the Technical in Participatory River Basin Governance

This chapter has taken a rather gloomy look at IWRM as a participatory governance framework in the Mekong River Basin. On the one hand, the several and often sincerely intended attempts to "do" participation by large scale organisations that subscribe to IWRM as the framework for integrating multiple objectives in transboundary river basin management in the Mekong have had little success in empowering farmers and fishers to manage the Mekong for their own well-being. On the other, the various participatory initiatives that challenge mainstream development thinking or that have responded to actual or potential impositions on local livelihoods have had limited success, and have tended to be more successful in single-issue campaigns than in institutionalising influence through an IWRM framework.

The message that this review of IWRM with respect to participation from above and below holds for the Mekong is, I suggest, twofold. First, we need to understand participation and IWRM as essentially political, and not merely as technical, and to recognise the systemic incapacity for the organisations responsible for implementing them to deal with them as such. Second, we need to understand the problems and issues that participation and IWRM deal with first and foremost from the perspective of those who live in the basin as part of their everyday existence, and not in terms of the projects that development agencies design to improve or otherwise impact on that existence.

Tania Li has written of the tendency of development projects, through their "will to improve", to "render technical" not only engineering projects but also the field of governance (Li 2007). We can see this tendency permeating the fields of participation and IWRM as practiced by mainstream agencies. Yet, both these fields are inherently political, and both are embedded in cultural practices. The real world of social engagement is one of negotiated outcomes, not of contrived platforms for orchestrated decision making. As a field of negotiation, participation remains a relevant category, for example in cases such as the 3SPN articulation of grievances and negotiation of restitution and shaming of perpetrators of injustice. Negotiated outcomes are also a sine qua non for societally embedded IWRM, given the implausibility of achieving a consensus-based approach to basin planning of the sort envisaged through MRC stakeholder forums such as those described above.

From the perspective of many localities in the Mekong River Basin, and despite the tendency of media, NGOs and other observers to see or portray critical threats (Osborne 2009), the Mekong does not seem to be a river system in crisis or requiring day to day "management", whether it be at sub-basin or whole-of-basin level. This is either because for the time being people continue to derive immense benefits from fisheries, river bank farming and the many other contributions the river makes to their livelihoods, or because the crises or hardships they do face do not present themselves as a system-wide phenomenon. The fishers of Tonle Sap whose catches have declined over past several years, the Sesan riverside communities in Ratanakiri whose bathing and drinking water has turned foul, the villagers near Pak Mun who have lost land and fishing-based livelihoods, and the people living along the Hinboun River in Laos all have explanations for what has gone wrong, but if those explanations go beyond the local, they tend to focus on specific scheme or agency targets rather than on structures of transboundary governance. Participation through engagement with those targets is most forceful and encompassing of affected people when it is based on clear sense of grievance.

We talk of IWRM as a framework for bottom-up approaches in a transboundary river basin, therefore, we need to be clear that we are not imagining some kind of representative body governed from below to manage the river and its basin. Rather, we are making the case for an integrative approach to management that does not leave local concerns behind, and one that recognises the inherently political and essentially critical nature of stakeholder engagement.

References

Arnstein S (1971) A ladder of citizen participation. J Am Inst Plann 35:216–224

Baird IG (2002) A community-based study of the downstream impacts of the Yali Falls dam along the Se San, Sre Pok and Sekong rivers in stung treng province, Northeast Cambodia, Se San Protection Network Project, Partners For Development (PFD), Non Timber Forest Products Project (NTFP), Se San District Agriculture, Fisheries and Forestry Office, Stung Treng District Office

Biswas A (2004) Integrated water resources management: a re-assessment. Water Int 29(2):248–256

Chou C (2010) The local governance of common pool resources: the case of irrigation water in Cambodia. Cambodian Development Resource Institute, Phnom Penh

Cohen PT, Pearson RE (1998) Communal irrigation, state and capital in the Chiang Mai valley. J Southeast Asian Stud 29(1):86–111

Cooke B, Kothari U (eds) (2001) Participation: the new tyrrany? Zed Books, London

Daravy K (2010) Challenges of participatory irrigation management in Cambodia: the case of Damnak Ampil irrigation scheme. MSc, School of Geosciences, University of Sydney, Sydney, p 183

Edmunds D, Wollenberg E (2001) A stretegic approach to multistakeholder negotiations. Dev Chang 32:231–253

Geertz C (1972) The wet and the dry: traditional irrigation in Bali and Morocco. Hum Ecol 1:34–39

Hirsch P (1997) Seeing forests for trees: environment and environmentalism in Thailand. Silkworm Books, Chiang Mai

Hirsch P (2008) 13 years of bad luck? a reflection on MRC and civil society in the Mekong. Watershed 12(3):38–43

Hirsch P (2010) Critique of draft basin development strategy. 3rd regional stakeholder forum. Vientiane, Lao PDR. Available at http://www.mrcmekong.org/download/3rd-stakeholder-forum/2.2-Philip-Critique-of-MRC-IWRM-Based-Basin-Dev-Strategy.pdf

Hirsch P, Lohmann L (1989) Contemporary politics of environment in Thailand. Asian Surv 29(4):439–451

Hirsch P, Wyatt A (2004) Negotiating local livelihoods: scales of conflict in the Se San River Basin. Asia Pac Viewpoint 45(1):51–68

Hirsch P, Jensen KM, Boer B, Carrard N, FitzGerald S, Lyster R (2006) National interests and transboundary water governance in the Mekong. Australian Mekong Resource Centre, School of Geosciences, University of Sydney in collaboration with Danida, Sydney

Lansing S (1987) Balinese 'water temples' and the management of irrigation. Am Anthropol 89(2):326–341

Li T (2007) The will to improve: governmentality, development and the practice of politics. Duke University Press, Durham

Molle F (2008) Nirvana concepts, storylines and policy models: insights from the water sector. Water Altern 1(1):131–156

Molle F, Floch P (2007) Water, poverty and the governance of megaprojects: the Thai "water grid". Unit for Social and Environmental Research, Chiang Mai University, M-POWER, Chiang Mai

MRC (Mekong River Commission) (2007) Independent, organisational, financial and institutional review of the Mekong river commission secretariat and the national Mekong committees: final report. Mekong River Commission, Vientiane, p 68

MRC (Mekong River Commission) (2010a) Draft minute: the 1st 7L sub-area meeting champasak grand hotel, Champasak province, 25–26 Feb 2010

MRC (Mekong River Commission) (2010b) Stakeholder analysis for the MRC basin development plan program phase 2: final report. Mekong River Commission, Vientiane, p 104

Osborne M (2009) The Mekong: river under threat, Lowy Institute Paper 27, Lowy Institute for International Policy, Sydney

Sheldon T (2005) River basin management: a negotiated approach. Both ENDS and Gomukh, Amsterdam

Thongdeelert C (1997) Cultural bioregionalism: towards a natural balance. Watershed People Forum Ecol 2(3):26–30

Chapter 9
The Dragon Upstream: China's Role in Lancang-Mekong Development

Darrin Magee

Abstract This chapter examines China's development on the upstream half of the Lancang-Mekong river, including the perspectives on local, regional, national, and international development that inform and motivate the nature and magnitude of that development. The primary goals of the chapter are to explain China's development approach to the Lancang-Mekong basin, Chinese development priorities for the upper half of the basin, and how those priorities are shaped and acted upon in China. I begin by describing the physical and human geographic characteristics of the Chinese half of the Lancang-Mekong Basin. Next, I lay out a series of problems or issues as perceived within China, and show that the corresponding solution to each problem wholly or partially justifies (from the Chinese development state's perspective) the construction of major infrastructure projects in southwestern and western China, of which the Lancang hydroelectric cascade is a major component. Finally, I discuss decision-making structures and practices in China and how they shape China's engagement with downstream neighbors regarding basin-wide development.

9.1 Introduction

Some 4,800 km upstream from the wide, muddy mouth of the Mekong River in Vietnam lies its source, a trickle fed by glacial runoff more than 5,000 m above sea level and known locally in Tibet as the Zaqu. From there, the Lancang (the river's Chinese name) tumbles down out of the eastern edge of the Qinghai-Tibet Plateau through Tibet and into Yunnan Province, through what one group of prominent Chinese Lancang-Mekong researchers calls the Longitudinal Range Gorge Region (LRGR) of China for its parallel, north-south-oriented river valleys (You et al. 2005).

D. Magee (✉)
Environmental Studies, Hobart and William Smith Colleges, Geneva, NY 14456, USA
e-mail: magee@hws.edu

J. Öjendal et al. (eds.), *Politics and Development in a Transboundary Watershed:*
The Case of the Lower Mekong Basin, DOI 10.1007/978-94-007-0476-3_9,
© Springer Science+Business Media B.V. 2012

At its upper reaches, the river is primarily nourished by glacial melt, and given the oft contentious politics of Southeast Asia that is the subject of much of this volume, it should come as no surprise that the very sediments in that melt-water are at the center of their own controversy.

From its source in Yushu County (Tibet), the Lancang drops some 4,700 m by the time it exits Yunnan Province, where it briefly forms the border between Laos and Myanmar (Burma) before flowing further south into Thailand, Laos, Cambodia, and Vietnam. This drop, along with the substantial volumes carried by the river, creates significant potential energy, which hydropower engineers and survey teams have eyed for nearly a century. Beginning with the development of the Manwan Dam in Yunnan in 1995, the first on the main stem of the Mekong, China's plans for a large-scale cascade of hydroelectric dams on the Lancang – especially on the 1,200-km stretch that lies within Yunnan – have been the catalyst for controversy over development within the entire Lancang-Mekong Basin. As a non-member of the Mekong River Commission, the uppermost riparian, and a longstanding regional power, China has become the easy target for criticisms over its "unilateral" development – perceived and real – of the river.

The Lancang watershed and surrounding areas, like most of China's western regions, have long been considered by easterners as backwards, undeveloped, and to a certain extent, undevelopable due to perceived deficiencies in the "quality" (*suzhi*) of the people there. Many residents are ethnic minorities assumed to lack the skills and education necessary to engage in the modern global economy, except perhaps as exotic objects of tourism. At the same time, the West's rich mineral and energy resources, including the massive hydropower potential of the middle and upper stretches of rivers such as the Lancang, Salween (Nu), and Yangtze (Jinsha), have created a development imperative to fuel energy- and resource-hungry urban and industrial centers in southern and eastern China. Over the past decade, discourses of poverty alleviation, rural electrification, and regional disparities have lent urgency and legitimacy to that imperative. More recently, China's emphasis on reducing the carbon-intensity of its economy by promoting a "low-carbon economy" (*ditan jingji*) has provided even more impetus to speed development of hydropower in the country's southwestern regions.

There are, of course, a number of other important economic activities within the Lancang watershed in China. Chief among these are extractive activities such as mining and farming; Yunnan has rich reserves of non-ferrous minerals, and its fruit and tobacco crops are prized throughout the country. Land-use surveys put roughly 79% of Yunnan's land area as agricultural land, with just over 2% used for development and construction, and the remaining 19% undeveloped (Wang 2002). Prior to the severe flooding on the Yangtze in 1998 that led to a ban on logging in the upper reaches of the river, forest products were also an important component of the provincial economy. Yet in terms of potential downstream impacts, the single most important form of economic activity taking place on the Lancang in China is hydropower development, the focus of this chapter.

9.2   Overview of the Lancang Cascade

Within China, the Lancang is itself generally divided into two (upstream and downstream) and sometimes three (lower, middle, and upper) sections. Almost all major development since the late 1980s has taken place on the downstream or lower portion, that is, the portion stretching from the site of the Gongguoqiao dam in west-central Yunnan to the point at which the river exits Yunnan (see Fig. 9.1). It is that development which has caused the greatest degree of concern among downstream countries, primarily due to fears that the massive reservoirs behind the Xiaowan and Nuozhadu dams, with their multi-seasonal regulating capacity, could fundamentally alter flow patterns downstream. Depending on the nature and magnitude of flow regime alterations, impacts to downstream fisheries, rice cultivation, navigation, and human livelihoods could result. In addition, uncertainty about the extent to which downstream water and sediment volumes are dependent on the upstream (Chinese) half of the river tends to further increase skepticism among downstream communities dependent on the river.

In addition to its hydropower resources, Yunnan also boasts abundant non-ferrous metal reserves and relatively large tracts of forested land, making extractive activities such as mining and logging important cornerstones of the provincial economy. While limited road networks throughout the more rugged parts of the province have connected those activities to larger economic centers in Yunnan and neighboring Sichuan, much of the central and western parts of northern Yunnan have remained relatively inaccessible (He and Wei 2008). Road construction intensified in the 1990s along the reach of the river north from Gongguoqiao to the Yunnan-Tibet border as interest in hydropower development grew, but development there has until recently been constrained by difficult terrain, limited infrastructure, and distance to load centers. As discussed below, this is now changing rapidly, facilitated by advances in transmission technology, a favorable policy environment and preferential investment priorities centered around energy resource development and the improvement in socioeconomic conditions presumed to follow. Huaneng Corporation, the largest of the five power generation corporations created from the assets of the former Ministry of Electric Power (MEP) and subsequent State Power Corporation of China (SPCC), has spearheaded the work through its Yunnan subsidiary Hydrolancang, which itself is a parent for more than a half-dozen subsidiary corporations responsible for individual dam development (Magee 2006b).[1] Other major parties include Sinohydro, an engineering and consulting group also created from the former MEP, as well as various consulting and construction subsidiaries of the Three Gorges Project Corporation.

---

[1] Hydrolancang is also known as Yunnan Huaneng Lancang River Hydropower Development Corporation. The Manwan dam was actually built by a partnership of the (then) Ministry of Water Resources and Electric Power, and the Yunnan Provincial Government. Similarly, Dachaoshan involved a consortium including Yunnan's famous tobacco group, Hongta, and three other state-run entities. The two dams are now owned and operated by Huaneng's principal Yunnan subsidiary, Hydrolancang. Much has been written elsewhere on reforms in the Chinese electric power sector since the mid-1990s (Magee 2006b; Xu 2002; Yeh and Lewis 2004).

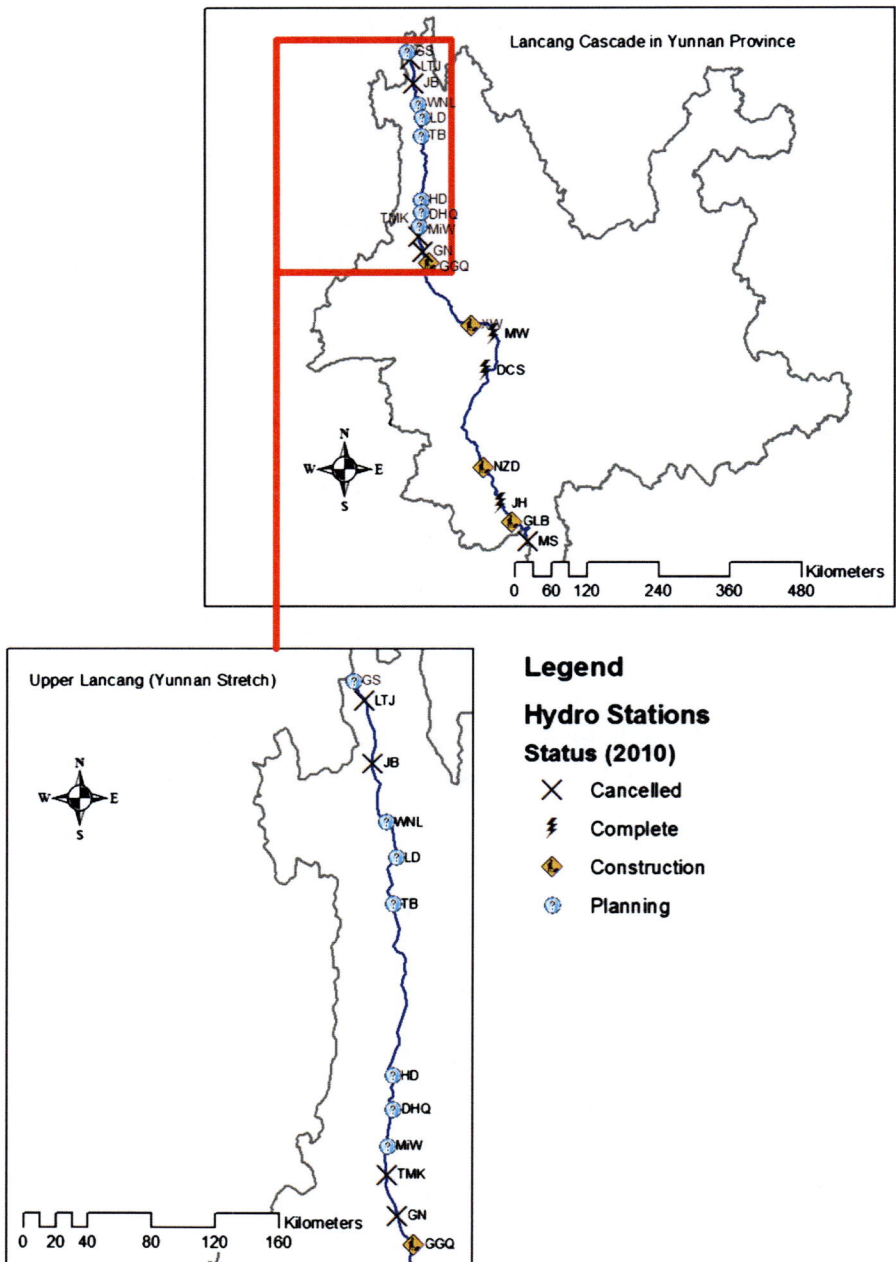

Fig. 9.1 Hydro stations in the Lancang, Yunnan province

Table 9.1 Details of lower Lancang hydroelectric cascade

Dam name	Installed capacity (MW)	Annual output (TWh)	Start date	End date	Dam height (m)	Reservoir volume (billion m³)	Map symbol
Gongguoqiao 功果桥	900	4.06	2006–2007	2015	130	0.51	GGQ
Xiaowan 小湾	4,200	18.89	2002	2012	292	15.13	XW
Manwan 漫湾	1,500	7.80	1986	1995	132	1.06	MW
Dachaoshan 大朝山	1,350	6.70	1997	2003	120.5	0.88	DCS
Nuozhadu 糯扎渡	5,850	23.68	2005	2017	260	22.74	NZD
Jinghong 景洪	1,750	7.93	2004	2011	107	1.23	JH
Ganlanba 橄榄坝	155	0.90		2015	60.5	0.072	GLB

Source: Magee (2006b) and various Chinese media and industry sources. Blanks indicate no reliable data were found

9.2.1 Lower Cascade

As shown in Table 9.1, the earliest dam completed on the lower reach of the Lancang (and indeed on the entire main-stem Lancang-Mekong) was the 1,550-MW Manwan Dam, which first came online in 1993 and was completed in 1995. Since then, work on the remaining six dams planned for the lower Lancang cascade has proceeded apace. The lower Lancang cascade is essentially a fait accompli; as of March 2011, three dams were completed (Manwan, Dachaoshan, and Jinghong); one was nearly completed (Xiaowan); and one was well underway and far too central to the cascade to be abandoned or significantly altered (Nuozhadu). The two smallest dams of the cascade (Gongguoqiao and Ganlanba) are now in the early stages of construction and will likely be completed by 2015. An eighth dam, Mengsong, was originally planned for the short stretch of the river between Ganlanba and the Myanmar border. With 600 MW of installed capacity, it would have been the second-smallest of the lower Lancang cascade in terms of generating capacity. According to a statement by China's Vice Foreign Minister Song Tao at the April 2010 Mekong River Commission Summit in Hua Hin, Thailand, the Mengsong dam has been cancelled due to concerns it would negatively impact fish migration through that stretch of the river ("China shows interest" 2010).[2]

---

[2]The statement by Vice Minister Song indicated that concerns about the impacts of unnatural fluctuations in water levels on fish migration were the primary cause for cancelling the Mengsong project. There is also some concern that the dam itself would impede fish passage. Even though the steepness, flow rate, and low temperature of the Lancang in more northern reaches make it inhospitable to migratory fish, studies by Chinese scientists suggest that at least some migratory species have traditionally made their way north of Jinghong.

One vital feature of the cascade-style development being pursued on the Lancang and its neighbors such as the Jinsha (Yangtze) and Nu (Salween) is that one or more dams in each cascade has an immense reservoir with multi-season regulating capacity. This translates into the ability of a reservoir to store water in significant quantities so as to ensure reliable power generation potential by that dam and downstream dams even during the driest months. On the lower Lancang, Xiaowan and Nuozhadu play that role. One industry publication estimated that the ability of Xiaowan to "bank" excess water during the rainy season and release it during the dry season will increase the power output of Manwan and Dachaoshan downstream by some 2 TWh annually (Zhu 2002).

While the large reservoirs at Nuozhadu and Xiaowan may be beneficial for power generation at those dams and their downstream counterparts, they are also easy targets for blame in the event of downstream water shortages. Most recently, as record low rainfall levels brought drought conditions to much of Mainland Southeast Asia in early 2010, many environmentalists claimed that the filling of the Xiaowan reservoir was causing the record low water levels in the Mekong (Ahuja 2010). The Prime Minister of Cambodia, as well as the head of the Mekong River Commission itself, both took pains to assert that lack of precipitation, and not the Chinese dams, were the root cause of the low water levels in the Mekong ("Chinese dams not to blame" 2010). The Chinese government, too, denied the accusations that the Lancang dams were the cause of the problem, arguing that southwestern China (including Yunnan) was at the mercy of the same drought conditions being felt in the lower Mekong watershed. The Chinese Ministry of Foreign Affairs went so far as to invite downstream officials on tours of the Lancang dams and offer daily flow data (heretofore considered a state secret). While the tours may have had some effect in allaying concerns, the sharing of daily flow data never seems to have materialized.

Perhaps in response to downstream concerns about altered flow regimes, the last dam on the Yunnan cascade, Ganlanba, is being designed and constructed to function as a counter-regulating (*fan tiaojie*) dam (Chai et al. 2007). If properly operated, Ganlanba could serve as a buffer between larger upstream dams and downstream water users, counteracting periods of low or high releases from those upstream dams by releasing more or less water, respectively. The end goal would be to mimic as closely as possible fluctuations in the natural flow regime over a variety of time scales, ranging from hourly and daily to weekly or even longer, while still generating power. Ganlanba, however, is the smallest dam in the lower Lancang cascade, and the ability of its relatively small reservoir to contain (or compensate for) greater volumes released (or withheld) by Jinghong dams would be limited to short time-frames. Moreover, in the report that followed approval of the Ganlanba pre-feasibility study (*yukexingxing yanjiu baogao*), no mention is made of the impact this re-regulating capacity on the cascade's flood control capabilities, which the developer and the Chinese government have often touted as a major benefit of the cascade to downstream countries. Instead, the only downstream benefit listed is improved navigation (Xishuangbanna Prefecture Development and Reform Commission 2008).

Table 9.2 Details of upper Lancang hydroelectric cascade

Dam name	Installed capacity (MW)	Annual output (TWh)	Dam height (m)	Reservoir volume (billion m^3)	Map symbol
Gushui 古水	2,600		305		GS
Liutongjiang (Cancelled) 溜筒江	550	3.45	130	0.5	LTJ
Jiabi (Cancelled) 佳碧	430	2.68	292	0.32	JB
Wunonglong 乌弄龙	960	4.366	136.5	0.265	WNL
Lidi 里底	420	1.952	74	0.075	LD
Tuoba 托巴	1,400	6.067	158	5.15	TB
Huangdeng 黄登	1,900	8.578	203	1.549	HD
Dahuaqiao 大华桥	800	9.79	106	0.23	DHQ
Miaowei 苗尾	1,400	6.468	139.8	0.722	MiW
Tiemenkan (Cancelled) 铁门坎	1,780	8.86	107	2.15	TMK
Guonian (Cancelled) 果念	1,200				GN

Source: Li et al. (2001) and various Chinese media and industry sources. Blanks indicate no reliable data were found

9.2.2   Upper Cascade

For the purposes of this chapter, "upper cascade" refers to the portion of the river within Yunnan and north of Gongguoqiao, as well as the principal upstream tributary of the Lancang, the Zaqu. In 2008, the Tibet Autonomous Region government (in particular, the TAR Development and Reform Commission) approved the preliminary plans for a stair-step cascade of dams on the Zaqu. The following year, Huaneng officially established its Tibet subsidiary, the Huaneng Upstream Lancang River Hydropower Corporation, to spearhead development on the Tibetan stretch of the river (Huaneng Corporation 2010). All the complexities of dam-building in northern Yunnan apply equally or to a greater degree in Tibet: limited roads and other infrastructure, challenging terrain, and long distances to load centers in the eastern part of the country.

Table 9.2 below provides basic information for dams that have at some time or another figured into plans for the upper Lancang within Yunnan Province. There is very little publicly available data on these dams, so this table should not be considered definitive. Some, such as Tiemenkan and Jiabi, do not seem to be part of current development plans, perhaps because the original site planned for the dam turned out to be unsuitable in terms of geology or proximity to UNESCO World Heritage sites in northern Yunnan (Fan 2005). Another, Guonian, was planned for northern Yunnan's Deqin County but reportedly cancelled due to concerns it might impact the already-receding Mingyong Glacier, even though those concerns were not yet scientifically validated (China Electric Power News 2007). Different sources – environmental impact reports, company press releases, media or government reports – frequently give different figures for key indicators such as installed capacity, reservoir volume, and annual power output, and more changes will likely come.

An analysis of various sources, though, suggests that somewhere around 10 GW of capacity will likely be built on the upper stretch of the Lancang within Yunnan.

Even less data is available on the dams on the Zaqu within the Tibet Autonomous Region. Huaneng's Tibet subsidiary claims a total of six dams will be built on the main stem of the Zaqu (Lancang); those, along with 100 MW on two tributaries, will provide a combined total of 6 GW of generating capacity for the Tibetan portion of the river (Huaneng Corporation 2010), slightly more than the capacity of the Nuozhadu station on the lower Lancang. As with the lower Lancang dams, those planned for the Tibetan reach of the river are billed as vital to alleviating local poverty and power shortages, and to furthering the integration of Tibet's economy with that of more developed eastern regions (Zhang 2010).

Aside from providing hydropower, the Lancang dams have also been touted as providing essential flood control services to downstream countries. Given the flooding that summer monsoon rains can bring to countries in the Lower Mekong basin, and the devastation such flooding can wreak on poor residents in low-lying areas, flood control provided by a string of Chinese dams would at first glance seem a good idea. Yet as I discuss below, flooding plays important ecological and socioeconomic functions in the Mekong watershed, and it is far from clear that the benefits of significantly curtailing that flooding would outweigh the costs. Moreover, the Lancang dams are multipurpose dams whose primary function is power generation. In southern China, demand for electric power peaks in summer, at the same time monsoon rains arrive and the need for flood control peaks. Put simply, hydropower generation demands releasing water while flood control requires storing it, making it essentially impossible to simultaneously maximize the two objectives with the same dam.

9.3   Chinese Perspectives on Lancang River Development

In this section I discuss the principal Chinese perspectives on Lancang development. While there is clear and demonstrated interest on the part of Chinese developers such as Sinohydro to gain a foothold in development projects on the Mekong outside China (especially in Southeast Asia and Africa), I focus here on development within China's boundaries. The section is organized around five issues that motivate hydropower development on rivers such as the Lancang. As I show, infrastructure development – with the Lancang hydroelectric cascade a central component – is an important part of the response to each issue raised, and has been pursued vigorously by actors within Yunnan, as well as national power companies, primarily the Huaneng Group and its Yunnan subsidiaries.

9.3.1   Domestic Electric Power Needs

China's demand for electric power has grown significantly in the past two decades, despite slight decreases in the rates of growth during the Asian financial crisis of

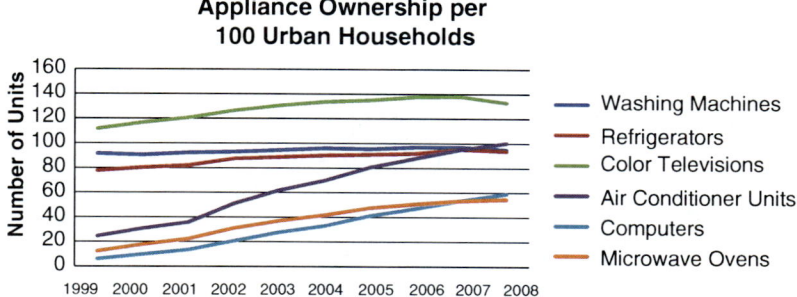

Fig. 9.2 Electrical appliance ownership per 100 urban households (Source: National Bureau of Statistics 2010)

the late 1990s, and the more recent global economic downturn that began in 2008. The primary driver of growth in demand for electric power has been the industrial sector, though household power consumption is becoming increasingly significant as disposable incomes rise and more Chinese households opt for creature comforts such as air conditioners and refrigerators. In southern and southeastern China, the heart of China's globally-oriented manufacturing engines, the annual demand curve peaks in summer when temperatures rise and the need to cool factories, warehouses, and office buildings is greatest. This is especially true in hot and humid Guangdong, where existing power generation infrastructure is already operating at its limits, and where rolling blackouts sent shocks through the provincial economy in 2003 (Fig. 9.2).

The blackouts served to intensify the search for a regional (sub-national) solution to power grid inadequacies. Policy slogans appeared that fit nicely under the broad "Western Development Campaign" (*xibu dakaifa*; see below), launched in 2000, naturalizing and legitimizing the link between power sources in western China and load centers in southeastern China: Send Western Electricity East (*xidian dongsong*); Send Yunnan Electricity to Guangdong (*Diandian Yuesong*); and, more recently, Send Tibetan Electricity Outward (*Zangdian waisong*) (Magee 2006b). China Southern Grid Corporation has invested heavily in new ultra-high-voltage trunk lines connecting the Lancang hydroelectric cascade, as well as a handful of coal-fired plants, to load centers across Guangdong.[3]

---

[3]Many of those lines are long-distance ultra-high-voltage direct current (UHV-DC) capable of transmission voltages up to ±800 kV. With these, China has become a world leader in UHV-DC technology, which is safer and less prone to transmission losses over long distances than alternating current (AC) lines. Given the distances – from several hundred to more than 1,000 km – involved in transmitting power from the Lancang dams to Guangdong, reducing line losses is a high priority.

9.3.2   Integration with Neighbors in the Greater Mekong Subregion

The electric power grid infrastructure throughout Mainland Southeast Asia is spotty, and swathes of Cambodia and Laos in particular lack reliable access to electric power. Many areas depend upon costly and highly-polluting diesel generators to feed local grids. Since the creation of the Greater Mekong Subregion (GMS) office of the Asian Development Bank in 1992, enhancing "subregional" infrastructure (including transportation, communication, and electric power networks) has been a key priority for member states (Asian Development Bank 2002). The newly created region consisted, oddly enough, of five countries (Vietnam, Cambodia, Laos, Thailand, and Myanmar) and one province (Yunnan), while notably omitting a second province-level entity in China that is home to the headwaters of the Mekong (the Tibetan Autonomous Region). It is important to note that at the time of the creation of the GMS in 1992, the status of organization currently known as the Mekong River Commission was uncertain; following the establishment in 1957 of the Mekong Consultative Committee for Coordinated Development, the organization had languished during the political upheaval and war that destabilized the region, during which Cambodia withdrew from the Committee. In 1991, Cambodia reapplied for membership, and the organization was reincarnated in 1995 (3 years after the establishment of the GMS office of the ADB) in its current form as the Mekong River Commission (Browder and Ortolano 2000).

Within the GMS, the Lancang-Mekong has the potential to serve a dual role, as both transport artery and power source. While growth in power demand across the GMS slowed following the Asian financial crisis, the argument that Yunnan can serve as a "battery" for (sub)regional power needs has great purchase within China (Yunnan Electric Power Network 2004). Indeed, one of the key policies shaping the direction (literally and figuratively) of electric power infrastructure in southwestern China is variously phrased as "Send Yunnan Electricity Outward" (*Diandian waisong*), which echoes the broader policy slogan of "Go Outwards" (*zou chu qu*), which the Chinese government has promoted heavily since the beginning of the 10th Five-Year Plan in 2000. As a first step in implementing that policy, power sales to Vietnam were initiated in 2004 via a 100-kV line crossing the border at Hekou/Lao Cai (Yunnan Electric Power Network 2004).

Thailand was identified as a promising market for Yunnan electricity in the late 1980s by planners at the Yunnan Electric Power Group (Yang 1998). An initial agreement between Hydrolancang and the Electricity Generating Authority of Thailand (EGAT) in the early 1990s called for Sino-Thai joint development of the Jinghong Dam. Following the Asian currency crisis, that agreement was renegotiated in favor of Thai investment in the development of both Jinghong and Nuozhadu, but the level of that investment is unclear. China Southern Grid has sought long-term power purchase agreements from EGAT that could involve the construction of a 500-kV transmission line from Yunnan to Thailand via Laos (Magee 2006a).

One major incentive for power sales across international boundaries has been the price, which is higher than that of power sold across provincial boundaries. An early contract between China Southern Power Grid and Electricité du Vietnam called for 1.3 billion kWh to be sold to Vietnam beginning in October 2006, for a total price of US $500 million (US $0.38, or roughly 3 Yuan per kWh, compared to the going rate within China of 1.4–1.8 Yuan per kWh) (Chen 2005).

9.3.3 Regional Disparities and Poverty Alleviation

The Western Development Campaign has shaped much of the internal development discourse in China from 2000 to the present, having been recommended by a central economic advisory commission in 1999 as the "major strategy for facing the new century" (Magee 2006a). Despite criticisms that the Campaign is misguided, aims to mollify areas perceived to have separationist tendencies (Wang 2004), or simply institutionalizes and legitimizes resource extraction patterns that had already existed (Goodman 2004), the Campaign likely also represents real concerns on the part of central government officials in terms of addressing economic development disparities between eastern and western China before they lead to widespread unrest (Naughton 2004). Priority areas under the Western Development Campaign have included investments in education, afforestation, new urban centers, and infrastructure such as railways, factories, and electric power facilities, especially hydropower (Magee 2006a). For hydroelectric projects, specific policies have included preferential loans and long-term repayment options (Zhang and Zhu 2001). As noted above, policies such as Send Western Electricity East and Send Yunnan Electricity to Guangdong make explicit the connections to be developed between Yunnan energy resources and eastern load centers.

The areas around the Lancang within Yunnan, where most of the hydropower development has occurred thus far, are quite poor and relatively inaccessible. Total population of the watershed within Yunnan is approximately five million people. The province covers 394,000 km^2 and is extremely rugged, with some 94% of its land area covered in mountains and elevations ranging from 6,740 m in the northwest down to 170 m in the southeast (Wang 2002). This topography, coupled with rivers descending sharply through steep, narrow gorges yields great hydroelectric potential, but also leaves many communities relatively isolated and with access to limited transportation and communications infrastructure. Many of the road networks that do exist have been carved out of hillsides to provide access to mining and logging operations. In this setting, large hydroelectric dams are billed as poverty alleviation projects through their ability to provide both employment and electricity to surrounding communities. In reality, however, this is not always the case: much of the construction skills needed for very large dams in difficult terrain are brought in via migrant construction labor, and the highly technical and automated nature of operating the dams once completed leave little room for a relatively low-skilled local labor force. Moreover, as noted above, the power generated from these dams is not distributed locally, but rather sent along dedicated high-voltage lines to load centers in provinces eastward.[4]

9.3.4 Low-Carbon Development

In recent years, as estimates of China's total carbon emissions approached and then surpassed those of the United States, until 2009 or so the world's number-one emitter of carbon, the Chinese government has begun promoting a "low-carbon economy" (*ditan jingji*) as one defining characteristic of its next phase of economic development. Indeed, "save energy and cut emissions" (*jieneng jianpai*) became an oft-repeated slogan in the latter years of the 11th Five-Year Plan (2006–2010), and progress on these goals will likely become an evaluation criterion for local officials in the 12th FYP. For much of the reform period (1978-present), the primary criteria for evaluation and promotion of officials have been their ability to maintain social order, attract investment (and grow local GDP), and meet family planning targets. Incentives to minimize pollution or energy use have essentially been nonexistent, and local officials often see tighter regulations on energy use or emissions as undercutting progress toward local economic development goals. Now, with low-carbon development a centerpiece of national policy, those incentives may change.

One way that the push to cut emissions has played out thus far has been through the permanent closure of old and inefficient coal-fired power plants, mostly in eastern China, and replacement with larger, more modern facilities whose emissions per kilowatt-hour are lower. The initial target set for the "close small, build big" during the 11th FYP in 2006 was to close 50 GW of such facilities by 2010. Reports claim that in actuality more than 70 GW (the equivalent of all of England's generating capacity) was permanently taken off-line during that period, a figure both provincial and central officials tout as testament to the country's commitment to cleaning up its development model (Jie 2010).

Yet China's electric power needs continue to grow, and alongside the construction of new, "cleaner" coal-fired power generation plants, a raft of hydroelectric, nuclear, wind and solar projects are being promoted as the low-carbon alternatives that will help power China's next phase of development. Large-scale hydropower is of fundamental importance here, and the extent of development planned for southwestern China – including the Lancang and neighboring Nu (Salween) and Jinsha (Yangtze) – cannot be underestimated. Hydropower currently provides

---

[4]To the extent the local, low-voltage power distribution grid in the vicinity of the dams is connected with higher-voltage transmission networks fed by the dams, one could argue that communities do indeed reap a marginal benefit from the dams in the form of increased stability and reliability of electric power.

roughly 15% of China's power output (National Bureau of Statistics 2010). China's theoretical hydropower potential is a staggering 676 GW (State Power Information Network 2008). Estimates of how much of that is actually technically and economically feasible vary over time and with technological advances; not surprisingly, the feasible figures have increased dramatically since the first surveys were conducted in the early 1900s, and now range from around 400 GW to 500 GW. Some 60% of that potential lies within the western regions of Yunnan, Guizhou, Sichuan, Chongqing, and Tibet (Zhou and Zhang 2003). The National Development and Reform Commission has set 380 GW as a development target for 2020. Given the number of projects currently underway or in the advanced planning stages, there is little reason to doubt that significant progress toward that target, if not the target itself, will be achieved by 2020.

9.3.5 *"Rational and Equitable Use" of Transboundary Waters*

As with any nation-state, the right of territorial sovereignty – and, by extension, the right to develop resources within one's own national boundaries – is seen as inviolable by the Chinese government. Indeed, it could probably be argued (though not here) that China is particular sensitive on matters of territorial sovereignty due to the so-called "Taiwan question," challenges to Chinese legitimacy in Tibet, border disputes with India, and competing claims to island groups such as the Spratley Islands and the Diaoyutai/Ryukyu Islands. Similarly, questions regarding development of natural resources (such as rivers for hydropower) and the importance of such projects for poverty alleviation in the Chinese countryside are, not surprisingly, seen as internal matters within China and therefore non-negotiable; the one caveat is that China claims to adhere to the principle of "rational and equitable use" of transboundary freshwater resources. As with many international law concepts, "rational and equitable use" is ambiguous at best, making it difficult for would-be discontents to mount a challenge to development within China's borders. Moreover, discourses of poverty alleviation are intertwined with policies promoting increased consumption of modern creature comforts in rural China such as refrigerators and air conditioners, all of which require greater reliability of electrical generation and distribution infrastructure.

The specifics, to the extent they exist, of "rational and equitable" in the Lancang-Mekong case have been explored to some extent already (He et al. 1999; Feng and He 1999). A more recent study (Feng and Magee 2009) found that, in terms of implementation, there may be some hope that institutions promoting cooperative management of transboundary waters can help mitigate hydropolitical tensions. Despite the fact that China is not a full member of the Mekong River Commission, ostensibly the most likely candidate institution for such cooperation, there is evidence from other transboundary basins within China that such institutions can promote coordination of development efforts. One example is the Tumen River Economic Development Programme, a joint project of China, South Korea,

Mongolia, and Russia.[5] Yet the particular geopolitical dynamics of the Lancang-Mekong watershed are fundamentally different, primarily due to the high levels of dependence on the Mekong of major segments of downstream countries' societies for livelihoods, along with longstanding suspicions among those countries of China's aims for regional hegemony and economic dominance.

Proponents of the Lancang dams have argued in the past that the dams will benefit downstream countries not only through increased and more reliable electric power provision, but also through flattening out the annual hydrograph of the river, moderately increasing dry-season flows while shaving the peak off of rainy-season floods (Liu 2010; He and Xiang 2004). While at first glance this seems a double-win for downstream countries, ecological science suggests that the peaks and valleys of annual high and low flows are vital for sediment transport, agricultural productivity, and regulating spawning activities of fish. Thus what is rational from a flood prevention and power provision perspective may be irrational from an ecosystem perspective.

9.4 China's Approach to Lancang-Mekong Transboundary Relations

The push to develop hydropower on the upper Mekong/lower Lancang over the past decade has coincided with increased attention among downstream countries, particularly Thailand, to the negative social and ecological impacts of large dams (Antaseeda 2002; Sneddon and Fox 2008). For instance, non-governmental organizations in Thailand and elsewhere have expressed concern that the experiences of villagers around the Pak Mun dam, on the Mun River near its confluence with the Mekong in northeastern Thailand, might be replicated elsewhere. They also frequently cite the ability of upstream dams in China, given the multi-seasonal regulation capacity of Xiaowan and Nuozhadu in particular, to fundamentally alter the high and low flows downstream (Goodman 2005). Such alterations could have adverse effects on riverbank stability (through sudden rises and falls of water levels on a daily time-scale) to fish spawning (through "leveling" of the annual hydrograph on a seasonal scale) (Gray 2002). The inter-governmental Mekong River Commission could, in principle, coordinate varying development priorities throughout the basin. China (along with Myanmar) presently maintains only observer status in the MRC. The effectiveness of the organization may be further constrained by given donor priorities that at times conflict with member country priorities, and the willingness of member states to craft bilateral agreements with each other and with China, such as an agreement between Laos and China on channel clearing in the Mekong (Gray 2002; Kumtita 2003).

---

[5]The project, supported by the United Nations Development Programme, is also known as the Greater Tumen Initiative. Information on the Initiative may be found on its website at http://www.tumenprogramme.org/

In recent years, China's relationship with the MRC has been somewhat schizophrenic. The Commission has at times defended China with surprising vigor, such as during the droughts of 2004 and 2010 which were first blamed on Chinese dams (Ahuja 2010; MRC 2004). Yet at other times it has complained publicly about China's lack of participation in the Commission and perceived nonchalance regarding the potential downstream impacts of upstream dam development (MRC 2002). With the release of its 2006–2010 strategic plan, the Commission embraced the potential for developing hydropower in the Mekong watershed, estimating that up to 30 GW could be developed on the main stem and major tributaries (MRC 2006). This perspective has been roundly criticized by environmental groups, many of which draw support from dam-affected community members throughout the four lower Mekong countries. Plans for some development on the Laos portion of the river continue to inch forward, but not without loud protests from those worried that damming the main stem in its lower reaches would devastate important fisheries such as Tonle Sap in Cambodia, which is estimated to produce fully half of the fish consumed in Cambodia (Middleton 2007).

Meanwhile, China has taken a more active role as a participant in the less political (and less politicized) Greater Mekong Subregion program. This is likely due to the emphasis of the GMS program on economic development, while skirting more politically sensitive issues such as data sharing and actually defining what "sustainable" development for the region (or subregion) might look like. In July 2005, the Second GMS Leaders' Summit was held in Kunming, the capital of Yunnan. The event was promoted with great fanfare and a makeover of much of the city, while a massive propaganda campaign was instituted on billboards, buses, and in newspapers feting the event and extolling the virtues of subregional cooperation. During the closed-door meeting, officials signed agreements furthering cooperation on subregional electrical grid development and interconnection, facilitating trans-border movement of people and goods, and preventing animal-borne diseases ("GMS huiyi" 2005).

As always, though, the effectiveness of intergovernmental instruments such as the MRC and GMS is tempered by the extent to which member states are willing to skirt the organizations and enter into bilateral agreements that do not necessarily align with the goals of the organization. One example of this was the decisions in the 1990s of the Electricity Generating Authority of Thailand to jointly invest in the Lancang dams (Jinghong, and later Jinghong and Nuozhadu)("Thailand Eyes Hydropower" 2002; "China, Thailand Inked Deal" 2000). A second was the agreement between China and Laos to clear rapids in certain stretches of the upper Mekong using dynamite (Gray 2002). While clearing rapids may aid navigation, it can also be detrimental to fish populations since rapids, and the rocks that create them, provide important habitat for fish. It is likely that a China-Laos proposal to remove rapids would have met with opposition had it gone through MRC channels, which could explain why Laos chose to enter into an agreement directly with China without going through the MRC. On a larger scale, the increasing level of involvement of Chinese firms in hydropower survey, planning, investment, and development in Mainland Southeast Asia, including on Mekong tributaries, is likely proceeding largely outside the ambit of the MRC given China's lack of membership status in the organization.

9.5 Hydropower Decision-Making in China

Decision-making regarding development on transboundary rivers in China is far from a straightforward process, and the case of the Lancang-Mekong is no exception. Aside from the competing interests of various ministries within China whose development priorities for the watershed may differ – shipping or forestry interests, for instance, may be at odds with hydropower development interests – the State Council, National Development and Reform Commission, and Ministry of Foreign Affairs all play important roles. Somewhat surprisingly (and perhaps clear evidence of the bureaucratic complexity of the Chinese state), the role of the Ministry of Water Resources in major hydroelectric projects, especially those on transboundary rivers, is rather ambiguous. The decision-making process is further complicated by the Ministry of Environmental Protection, China's basin (or watershed) commissions, and the quasi-governmental status of China's major energy development corporations. This section briefly examines each of these agencies and their role in decisions on Lancang development.

The State Council is chaired by the Premier, made of up heads of central government ministries, commissions, departments and agencies and numbers about 50 people total. As the highest executive body within the Chinese state, it holds final decision-making authority on a wide variety of topics, and bases such decisions at least partly on reports and assessments from the various ministries and ministry-level organizations involved. For hydropower, the State Council pays particular attention to large-scale projects (usually those with greater than 250 MW of installed capacity) and projects on international rivers. The point at which the State Council intervenes in the decision process for any one project seems to vary, though by virtue of its status, it can essentially approve or veto any project. Aside from international rivers, the State Council takes direct responsibility for all other rivers considered of primary importance (*zhuyao*).

The NDRC (formerly the State Development and Planning Commission) is generally recognized as the primary force behind large-scale development projects of all sorts in China. It also plays a coordinating role, balancing the priorities of various ministries. The Commission houses an Energy Office that is responsible for researching domestic and international energy development, proposing strategies and plans for domestic energy development along with suggestions for related institutional reforms. It is also responsible for managing petroleum, natural gas, coal, and hydropower resources, though it is not clear how this management overlaps with other agencies entrusted with the same responsibilities. The NDRC Office of Comprehensive Utilization of Environment and Resources is charged with studying ways to resolve conflicts between economic development, social development, and environment. This office is responsible for environmental planning, including resource conservation planning and green production methods. Perhaps the most visible impact of the NDRC on economic development are the Five Year Plans, which set broad development goals and directions for the economy that are then interpreted and implemented by ministries, local governments, and corporations.

The Ministry of Water Resources has in the past been repeatedly combined with and separated from various incarnations of the Ministry of Electric Power and/or Hydropower. Currently, the MWR houses a Department of Planning and Programming that is theoretically responsible, among other things, for integrated river basin planning as well as siting and storage planning for large- and medium-scale hydropower stations. In addition, the Ministry's Department of Construction and Management is responsible for "giving guidance" and overseeing safety of reservoirs and dams for water supply and hydropower development. Overall, however, the MWR tends to focus on all aspects of water resources except hydropower, and it only gets involved in hydropower projects to the extent they have an impact on sedimentation, navigation, irrigation, or other non-power uses of water resources.

The Ministry of Environmental Protection, formerly the State Environmental Protection Administration, has only recently been promoted to Ministry status. This is important in China, where sensitivity to the administrative rank of institutions and individuals is high. The administrative promotion, however, has not been accompanied by significant increases in staffing or budget, which limits the MEP's effectiveness. Moreover, with regard to water resources, the MEP is primarily charged with policing water quality, an increasingly daunting task given the widespread and rapid nature of industrial development in China, coupled with the lack of industrial and municipal wastewater treatment plants in much of the country. If anything, the MEP tends to take a supportive stance on hydropower, given nationwide pressures to reduce emissions and hydropower's broadly accepted status in China as a clean energy source.

China's seven basin commissions, originally tasked with designing and carrying out water conservancy projects throughout China's principal watersheds, have grown into major bureaucracies charged with the daunting task of "comprehensive development" along all of China's major rivers. The largest of these, both in terms of watershed and manpower, is the Yangtze River Commission.[6] Based in the southern, middle Yangtze city of Wuhan but with branch offices throughout the watershed, the Commission boasts some 30,000 employees. In addition to the Yangtze watershed, it is also tasked with overseeing planning and coordinating development for the Lancang, Nu, and other southwestern rivers. Interestingly, the Yangtze River Watershed Commission is actually a delegated agency (*paichu jigou*) of the State Council. As such, the Commission, in principle at least, holds the authority to approve or reject projects based on whether or not they meet the requirements of the comprehensive plan for a particular river basin. Interviews with hydropower developers and Commission officials, however, suggest that the authority of the Commission in this regard is far from

---

[6]The other commissions are the Yellow River Water Resources Commission, Huaihe River Water Resources Commission, Haihe River Water Resources Commission, Pearl River Water Resources Commission, Songliao River Water Resources Commission, and Taihu River Water Resources Commission.

clear, and that ambiguities arising from the Commissions delegated status with respect to the State Council, and its subordinate status to the Ministry of Water Resources, muddies the waters in terms of decision authority.

A final set of actors that play a key role in hydropower development decisions in China are the major power corporations themselves, all the progeny of the former Ministry of Electric Power. In fact, there is compelling evidence that much of the so-called comprehensive development planning for some key rivers, including the Lancang, is being spearheaded not by the Yangtze River Basin Commission, but by the Huaneng Corporation, the largest of China's electric power conglomerates. Interviews with company representatives and with officials from the basin commission and Ministry of Water Resources all suggest that the ability of hydropower interests to fundamentally shape the overall development of particular rivers derives at least in part from the former-ministry status of those corporations, and the preferential access that status affords them to decision-makers in central government bodies such as the State Council and NDRC.[7]

The Ministry of Electric Power was reorganized as the State Power Corporation of China in 1998, which was then divided into five generation companies, two grid companies, and four consulting and design companies in 2002 (Xu 2002; Yeh and Lewis 2004; Magee 2006b). This restructuring occurred in a context of broader reforms in the state-owned enterprise (SOE) sector and amid calls to separate generation from transmission (*changwang fenkai*) in order to promote greater competition in the power sector. Huaneng, which inherited the lion's share of generation assets, was granted development rights for the Lancang by the State Council, which, as noted above, it has carried out through its subsidiaries Hydrolancang and Huaneng Upstream Lancang River Hydropower Corporation. Similarly, most of the design work for the Lancang dams has been carried out by subsidiaries of Sinohydro and Hydrochina, two of the four planning and design entities formed from the former Ministry of Electric Power.

The actors detailed here represent an important subset, not an exhaustive list, of all the interests that influence development decisions for the Lancang River. Obvious omissions here include various offices within local governments at the provincial level and below, as well as academic institutes, non-governmental organizations, and representatives of affected communities. The Communist Party, of course, maintains party committees in all government offices and within important companies (SOEs or quasi-SOEs), and can sway decisions in one direction or another according to particular political priorities. Given the broad constellation of actors involved in development decisions, and the often vague language of China's administrative laws (such as that governing the role of the basin commissions vis-à-vis the Ministry of Water Resources), it should come as no surprise that one's view of decision-making processes on river basin development may depend strongly on where one stands within that process. In the Lancang case, tensions between those competing visions arise primarily from different understandings of the role of the Yangtze River Basin

---

[7]Aside from China Huaneng, the other four generation corporations are China Datang, China Huadian, China Guodian, and China Power Investment.

Commission and the Huaneng Corporation and its subsidiaries. The perspective of the former is that comprehensive planning, led by the Commission, should precede hydropower development, and that individual hydropower projects should be approved by the Commission (as the delegated agency of the State Council). Interviews with Commission officials and Huaneng representatives, however, suggest that in actuality the developer is apt to take a more direct route, seeking approval for projects directly from the State Council, skirting the authority of the Commission.

9.6   Conclusion: Conflict and Commonality in Development Priorities

In this chapter I have sought to analyze China's role in Lancang-Mekong development. Since hydropower on the main stem of the Lancang-Mekong is the most contentious and contested form of development, I have concentrated my analysis on dam development processes on the Chinese stretch of the river, ignoring for the moment the role of Chinese corporations in downstream development outside China. In concluding this chapter, it seems important to step back from the intricacies of decision processes in China to ask a few larger questions about China's role as upstream riparian and regional power. First, how do development priorities in China conflict with those of the downstream riparians? Second, what do Chinese concerns about sovereignty and national security mean for Integrated Water Resource Management (IWRM) goals in the Lancang-Mekong basin? And finally, to what extent might there be increasing alignment between Chinese and MRC/lower riparian goals for the Lancang-Mekong?

Chinese development of hydropower resources on the Lancang began a quarter-century ago in the mid-1980s, with detailed planning for such development having begun even earlier. Downstream groups outside China concerned about environmental and social impacts grew increasingly vocal in their opposition throughout the late 1990s, with the publication of the World Commission on Dams report (2000) lending further credibility to their complaints. Within China, the development of the Lancang dam cascade has been quite effectively legitimized through discourses of western region development, poverty alleviation, energy security, flood control, and sustainable (and "scientific")[8] development. More recently, concern about emissions reduction has lent further momentum to hydropower development as a low-carbon alternative to coal-fired power generation.[9] While

---

[8]The term "scientific development perspective" (*kexue fazhan guan*) came into broad use in development and policy circles in China during the 11th Five-Year Plan (2006–2010), and often seems to be conflated with sustainable development.

[9]It is important to note, though, that hydropower's share of overall power generation is not expected to grow appreciably in the coming decades, as investment in new, larger coal-fired plants will likely continue for the foreseeable future.

none of these priorities, except perhaps emissions reductions, is likely to immediately benefit downstream countries, the extent to which they might harm downstream social and ecological communities is unclear.

The perception that China has opted to "go it alone" on Lancang development has been reinforced by the country's continued low level of engagement with the MRC and foot-dragging on data-sharing initiatives. Indeed, much of the concern about potential downstream impacts such as bank erosion, fish migration and habitat destruction, and land subsidence might well be alleviated through greater data transparency on indicators such as daily flow volumes and sediment transport. Yet the tendency in China is to assume that data that are not explicitly "open" (*gongkai*) should be considered as *not* open and shared with extreme caution if at all, even if they are not explicitly considered "internal" (*neibu*, or classified) either. Such caution is understandable; given the Chinese government's propensity to occasionally crack down on activists or even academics for doing anything considered threatening to national security, there is frankly little incentive for Chinese scholars, who may expend considerable political capital to acquire that data through personal connections and back-door channels, to share that data with organizations or individuals outside China.

Finally, while China and its Mekong neighbors may be in agreement on broad development goals such as subregional poverty alleviation, improved transportation infrastructure, and more reliable access to energy resources, the means for achieving those goals may differ. So, too, does the ability of disenfranchised communities to gain access to decision processes regarding those development initiatives, even (or perhaps especially) in cases where those communities stand to benefit least from megaprojects such as hydroelectric dams and high-speed roadways. Given the apparent increase in convergence of Chinese and MRC priorities regarding mainstem hydropower, a prospect welcomed by Chinese hydropower developers already involved in survey and design work throughout the GMS, it is vital for scholars, activists, and non-governmental organizations to continue to press for greater data sharing (in both directions) and transparency in decision-making processes regarding development in the basin.

Acknowledgments This work was partly supported by funds from the US National Science Foundation Grant BCS-0826771. The author wishes to thank Ms. Congjing Zhong, an undergraduate research assistant at Hobart and William Smith Colleges, for her careful assistance with Table 9.2. Any errors remain my own.

References

Ahuja A (2010) China says dams not to blame for low Mekong levels. http://www.reuters.com/article/2010/04/05/us-mekong-idUSTRE6341A620100405. Accessed 28 Feb 2011
Antaseeda P (2002) Upstream Power Play. http://www.livingriversiam.org/mk/Mek_dam_nE3.htm. Accessed 7 Sep 2011
Asian Development Bank (2002) Building on success: a strategic framework for the next ten years of the greater Mekong subregion economic cooperation program. Asian Development Bank, Manila

Browder G, Ortolano L (2000) The evolution of an international water resources management regime in the Mekong river basin. Nat Resour J 40(3):499–531

Chai Y-S, Pan C-Y, Li L-Y (2007) Ganlanba shuidianzhan bazhi, baxian xuanze (Selection of dam site and dam line of ganlanba hydropower station). Yunnan Shuidian Jishu 160:29–32

Chen H (2005) Shijie shou tiao dianli 'chaoji gaosu' you Dian beng Yue, mingnian kaijian (First 'ultra-high-speed' electricity galloping from Yunnan to Guangdong, work begins next year). http://www.pprd.org.cn/yunnan/hezuo/200510130011.htm. Accessed 17 Apr 2006

China Electric Power News (2007) Zhongguo Dianli Bao: Lancang Jiang de lvse suanshu (China electric power news: lancang river's green math). http://www.chng.com.cn/n31531/n31605/c355589/content.html. Accessed 3 Apr 2011

China shows interest in taking MRC membership (2010) http://www.bangkokpost.com/news/local/35637/. Accessed 1 Mar 2011

Chinas Thailand inked deal to tap Chinese hydro-power (2000) http://fpeng.peopledaily.com.cn/200005/03/eng20000503_40142.html. Accessed 13 Nov 2003

Chinese dams not to blame for low Mekong levels: Cambodia PM (2010). http://www.terradaily.com/reports/Chinese_dams_not_to_blame_for_low_Mekong_levels_Cambodia_PM_999.html. Accessed 1 Mar 2011

Fan X (2005) Shuidian da kaifa dui shijie yichan yiji guojia fengjing lüyoudi de yingxiang (The influence of great hydropower development on World Heritage and National Scenic areas). http://www.fjms1984.com.cn/20040101/ca689550.htm. Accessed 1 Mar 2006

Feng Y, He D (1999) Guoji shuifa de fazhan qushi yu Zhongguo shuifa tixi de duibi fenxi yanjiu (A comparative study on the developing tendencies of international water law and Chinese water law system). Dili Xuebao Acta Geogr Sinica 54(Suppl):165–172

Feng Y, Magee D (2009) Hydropolitical vulnerability and resilience in international river basins in China. United Nations Environment Programme, Nairobi

GMS huiyi jiang qianshu liu da wenjian (2005) Six major documents to be signed at GMS meeting. http://www.yndaily.com/html/20050703/news_86_210705.html. Accessed 3 July 2005

Goodman DSG (2004) The campaign to 'open up the west': national, provincial-level and local perspectives. China Quart 158:317–334

Goodman PS (2005) Manipulating the Mekong. Washington Post, p. E01. http://www.washingtonpost.com/wp-dyn/articles/A35167-2004Dec29.html. Accessed 10 Sep 2011

Gray D (2002) Chinese dams, channel blasting may spell disaster for mighty Mekong River, environmentalists say. Associated Press. Available via LexisNexis

He S, Wei Z (2008) Lancang Jiang shangyou ge tiji shuidianzhan duiwai jiaotong yunshu fang'an xuanze (Selection of transport alternatives for various cascaded power stations on upper reaches of Lancang River). Yunnan Shuidian Jishu (Yunnan Hydropower Technology) 162

He G, Xiang C (2004) Wo guo zhongdian shuidian jianshe xiangmu Xiaowan dianzhan dajiang jieliu chenggong (River successfully blocked at China's key hydroelectric construction project, Xiaowan). http://news.sina.com.cn/o/2004-10-26/10564043086s.shtml. Accessed 22 Apr 2011

He D, Yang M, Feng Y (1999) Xi'nan guoji heliu shuiziyuan de heli liyong yu guoji hezuo yanjiu (Study on reasonable utilization of water resources in international rivers and international region cooperation in Southwest China). Dili Xuebao Acta Geogr Sinica 54(Suppl):29–37

Huaneng Corporation (2010) Huaneng Upstream Lancang River Hydropower Co., Ltd. http://www.chng.com.cn/n93521/n93834/c95347/content.html. Accessed 1 Mar 2011

Jie Z (2010) Zhongguo wunian taotai xiaohuodian jizu guimo chaoguo Yingguo huodian zhuangji rongliang (Thermal power capacity retired by China in five years exceeds England's total installed capacity). http://www.in-en.com/power/html/power-1429142951771331.html. Accessed 26 Oct 2010

Kumtita T (2003) Mekong Reef Blasting: Review of Project Demanded. Bangkok Post. http://www.ecologyasia.com/news-archives/2003/jan-03/bangkokpost_030110_1.htm. Accessed 15 Sep 2011

Li Y, Tang X, Li P, Qin M, Ji H, Ma S (eds) (2001) Da xi'nan yu Lancang Jiang – Meigong He ciquyu hezuo kaifa (The great southwest and Lancang-Mekong subregional cooperative development). Yunnan minzu chubanshe, Kunming

Liu F (2010) Xiaowan shuidianzhan fanghong duxun guanli shijian (The practice of flood control and tiding-over management of the Xiaowan hydropower plant). Yunnan Shuili Fadian 26(6). doi:10.3969/j.issn.1006-3951.2010.06.011

Magee D (2006a) New energy geographies: powershed politics and hydropower decision making in Yunnan, China. Phd dissertation, University of Washington, Seattle

Magee D (2006b) Powershed politics: hydropower and interprovincial relations under great western development. China Quart 185:23–41

Middleton C (2007) Protecting the fisheries of Tonle Sap lake. http://www.internationalrivers.org/southeast-asia/mekong-mainstream-dams/protecting-fisheries-tonle-sap-lake-0. Accessed 20 Apr 2011

MRC (Mekong River Commission) (2002) Closer cooperation with China needed, says Mekong River Commission. http://www.mrcmekong.org/media/press2002/press005.htm. Accessed 20 Oct 2002

MRC (Mekong River Commission) (2004) Mekong's low flows linked to drought, says MRC study. http://www.mrcmekong.org/MRC_news/press04/26-mar-04.htm. Accessed 27 Mar 2004

MRC (Mekong River Commission) (2006) MRC Strategic Plan 2006–2010. http://www.mrcmekong.org/annual_report/2006/strategic-plan.htm. Accessed 12 Jan 2007

National Bureau of Statistics (ed) (2010) Zhongguo tongji nianjian (China statistical yearbook) (China statistical yearbook). Zhongguo tongji chubanshe, Beijing

Naughton B (2004) The western development program. In: Naughton B, Yang D (eds) Holding China together: diversity and national integration in the Post-Deng era. Cambridge University Press, New York, pp 253–296

Sneddon C, Fox CA (2008) Struggles over dams as struggles for justice: The World Commission on Dams (WCD) and anti-dam campaigns in Thailand and Mozambique. Society and Natural Resources 21(7):625–640

State Power Information Network (2008) Zhongguo shuineng ziyuan gaikuang (Overview of China's hydropower resources). http://www.sp.com.cn/zgsd/zgsygk/200805/t20080515_104282.htm. Accessed 15 Apr 2011

Thailand Eyes Hydropower Project in Southwest China (2002) http://china.org.cn/english/scitech/37159.htm. Accessed 13 Sep 2011

Wang S (ed) (2002) Yunnan Dili (Yunnan geography). Yunnan minzu chubanshe, Kunming

Wang S (2004) The political logic of fiscal transfers in China. In: Lu D, Neilson WAW (eds) China's west region development: domestic strategies and global implications. World Scientific Publishing Co., Inc, River Edge, pp 101–136

World Commission on Dams (2000) Dams and development: a new framework for decision-making. Earthscan, London

Xishuangbanna Prefecture Development and Reform Commission (2008) Ganlanba shuidianzhan yukexingxing yanjiu baogao tongguo shencha (Ganlanba Hydropower Station Pre-feasbility Study Report Approved). http://dl.xxgk.yn.gov.cn/canton_model3/newsview.aspx?id=396429. Accessed 1 Dec 2009

Xu Y-C (2002) Powering China: reforming the electric power industry in China. Ashgate, Aldershot

Yang R (1998) Jiakuai Xiaowan shuidianzhan xiangmu de choujian cujin ziyuan youhua peizhi (Hasten preparations for the xiaowan hydropower project, promote optimal resource allocation). In: Zhongguo kexue jishu xiehui xuehui bu (ed) Zhongguo xibu nengyuan ziyuan kaifa ji youhua peizhi xueshu yantaohui wenji. Sichuan Science and Technology Press, Chengdu, pp 139–145

Yeh ET, Lewis JI (2004) State power and the logic of reform in China's electricity sector. Pac Aff 77(3):437–465

You W, He D, Duan C (2005) Yunnan zongxiang linggu diqu qihou bianhua dui heliu jingliu liang de yingxiang (Climate change of the longitudinal range-gorge in Yunnan and its influence on the river flow). Dili Xuebao Acta Geogr Sinica 60(1):95–105

Yunnan Electric Power Network (2004) Woguo shouci daguimo xiang guowai mai dian (China's first large-scale electricity exports). http://www.yndl.com/shownews.asp?newsid=1578. Accessed 27 Sept 2004

Zhang L (2010) Gongsi chengjian Xizang Zaqu Guoduo shuidianzhan gongcheng (Company [Sinohydro Bureau 5] Takes Tibet Zaqu Guoduo Hydropower Project). http://www.zswjwfj.com/htm/1/2/177.html. Accessed 15 Mar 2011

Zhang X, Zhu W (2001) Xibu da kaifa: jingji, tongji, juece (China's western development: economics, statistics, and decision-making). Jingji guanli chubanshe, Beijing

Zhou D, Zhang J (eds) (2003) Zhongguo shuili fadian nianjian 2001–2002 (Almanac of China's water power 2001–2002), vol 7. Zhongguo dianli chubanshe, Beijing

Zhu T (ed) (2002) 20 Shiji heliu shuidian guihua (20th century river hydropower planning). China Electric Power Publishers, Beijing

Chapter 10
Politics vs Development in a Transboundary River Basin – The Case of the Mekong Basin

Joakim Öjendal, Stina Hansson, and Sofie Hellberg

Abstract This concluding chapter looks at the nexus of development and politics in a transboundary basin by bringing together the different contributions to the volume and their different approaches to this problematique. Together they help us unpack the perceived incompatibility between an integrative participatory approach and international dynamics relying on a sovereignty logic in a contested space. Four central themes are discussed, namely the limitation of institutions and agreements; the securitization and national prerogative in water management; the tendency of IWRM approaches to hide politics and allow business as usual; the politics of participation and the challenges to its implementation; and finally knowledge management as a key to make a difference – given that its inevitable politicization is taken into account. Taking a critical stand the chapter still emphasizes the possibility to alter politics, on condition of its recognition. The challenges and dilemmas should not be used as an argument for not continuing to work with the politics of water management.

This book sets out to address the urgency and complexity of efficient water governance and transboundary cooperation. As has been convincingly established through empirical research (Wolf 1997), theoretical reflections (Mirumachi 2010), as well as through scrutiny of practitioners' approaches (Earle et al. 2010), it is neither the fear of violence nor that of overall macro-political conflicts that constitutes the dominant impediment to sound transboundary water management. Rather it is the *quality* and *depth* of attempts at cooperation and of development that is the weakest link towards enhanced basin-based water governance. That is, how are the mechanisms of proclaimed cooperation functioning and with what effects on proposed development and its compatibility with the transboundary limitations?

To a varying degree, national and international water resources management is adhering to Integrated Water Resource Management (IWRM) principles with focus

J. Öjendal (✉) • S. Hansson • S. Hellberg
School of Global Studies, University of Gothenburg, Box 700, 405 30 Gothenburg, Sweden
e-mail: Joakim.ojendal@globalstudies.gu.se; Stina.hansson@globalstudies.gu.se;
Sofie.hellberg@globalstudies.gu.se

J. Öjendal et al. (eds.), *Politics and Development in a Transboundary Watershed:*
The Case of the Lower Mekong Basin, DOI 10.1007/978-94-007-0476-3_10,
© Springer Science+Business Media B.V. 2012

on participation, integration and finding solutions for a sustainable management of water resources. As such IWRM has become the dominant approach to water management on all levels. However, IWRM has been criticised for being merely a rhetorical device allowing for business as usual, for keeping a status quo rather than finding ways of realizing progressive solutions to substantial problems of water management, and for having a tendency to nullify politics, and to wish away power and interests (i.e. hiding contradictions instead of solving issues).

However, as we see it, as a diverse set of interests and complex power dynamics are inherent in issues of water allocation, to evade politics and power is an impossibility. An unwillingness to deal with them will rather result in hiding politics from view, which then results in the opposite of what the IWRM approach preaches (intransparency, disintegration, power politics). Moreover, in a transboundary context issues of politics and power risk to kick in with even sharper intensity and to contribute to securitization of water management, which could involve a furthering away from sustainable, inclusive, joint and/or participatory management of water. This is more thoroughly outlined in the Introductory chapter to this volume (Chap. 1).

Hence, the purpose of this book has been to put light on the assumedly incompatible logics of IWRM with its progressive values and practices on the one hand, and international/regional politics with a state logic, emphasising sovereignty and national interests on the other. Although harboring contrary values, these two processes – IWRM-based water governance approaches and regional politics – coincide, being mutually negotiated, and occasionally enjoying synergy and mutual stimulation. When, where, and how these different outcomes come into play is a scantily researched field. Therefore we set out to empirically scrutinize some of its articulations in the case of the Mekong in order to better understand the potential for sustainable solutions to water resources management in the region. This is of additional urgency since the basin is characterized by high poverty rates and rapid economic development that put severe strains on the Mekong water resources and may in case of crisis result in increasing tension in the region as well as failed attempts at utilizing valuable water resources in productive *and* sustainable ways. These are features that apply to many transboundary basins in and out of the developing world, rendering us to believe that the findings here have value beyond the Mekong.

Our contributors have all in different ways addressed this problematique. As we read them, looking at the nexus of development and politics in a transboundary basin, some issues emerge of a generic nature which we will illuminate below with the help of the empirical chapters.

10.1 Findings and Reflections

10.1.1 Agreements and Institutions – and Their Limitations

First of all, as Swain inspires us to believe, and what most contributors underline ex- or implicitly, basin agreements are not in themselves the solution to overcome the problems of IWRM in a transboundary context. Swain shows how agreements have staved off conflict and opened up the discussion of water peace, but he convincingly argues that they have still failed to produce constructive outcomes in terms of allocation and sustainable use of water resources. Moreover, it is argued that agreements have been inept at dealing with increasing water demand and adaptation to climate change. Actually, most water agreements and River Basin Organisations (RBOs) base their activity on very rigid water allocation formulas, which may prove to be contrary to flexibility and sustainable agreements when climate and water amounts start to change. As an irony of history, the 'failure' of the MRC-Agreement negotiation – that exact figures on water flow were never agreed on for wet and dry seasons and that allocation formulas were never concluded (cf. Chap. 3 Öjendal and Mørck Jensen) – may prove a blessing in an era of climate change: the rigidity coming with fixed numbers may be avoided in the Mekong case.

However, one of the main problems is that institution-building around river-basins are not sufficiently addressing the political nature of development and transboundary cooperation. Hirsch points at the general consensus in the literature and among practitioners that the MRC is rather weak in its role to govern water-related decisions. Three main causes are presented, the absence of China as a member, the consensus style politics that governs the organization, and the lack of embeddedness of the organization in the affairs of the member countries. In combination these go a long way to explain the governance weakness of the MRC. Moreover, when water management issues are captured by the higher echelons of power in the respective countries, the influence of the MRC and the national Mekong committees is further weakened. This is accentuated by the fact that MRC is often seen as donor driven and therefore lacking regional ownership.

As several of the contributions in this volume show, current basin cooperation allows national logics to prevail, and articulated 'cooperation' may in its extreme rather be seen as a trampoline to exercise a more full national agenda. For example, Mirumachi, in her contribution, shows how Thailand has simultaneously been able to engage in the cooperative framework *and* has managed to maximize capture, in spite of the MRC which basically exist in order to prevent that logic to prevail. To follow Mirumachi, in spite of a signed agreement and a crafty institution, cooperation remains elusive as it is permeated by politics and a state economic development imperative. Especially in a not-yet-closed basin as the Mekong, the 'regional' becomes a vehicle for negotiation on how to pursue these agendas, chasing to catch the still 'available' waters.

At some point we may need to recognize, as Swain and Hirsch for different reasons remind us, that there is a systemic incapacity of regional organizations to be responsible for implementing IWRM in a transboundary context if this is not supported by the states sharing the basin. Anything else would be to place unrealistic expectations with an actor unlikely to solve the structural problem of national sovereignty and water cooperation.

10.1.2 Securitization and the National Prerogative

It can be argued, as Cooper and Mirumachi separately do in this volume, that the status of river management as essential for the economic development of the state brings it to the higher echelons of state power, and thereby constitute a form of securitization, where water management is lifted from the remit of the national Mekong committees and water ministries to the ministries of finance, planning and foreign affairs. This position makes stakeholder dialogue more difficult as water management enters the field where national interest and proclaimed sovereignty kicks in. Securitization of water resources management has had the effect of legitimizing large-scale infrastructure development as well as limiting participation in decision-making processes, and thereby to legitimize certain types of water resource management that is contrary to the integrative, sustainable and participatory objectives of IWRM. As Öjendal and Mørck Jensen point out, the Xayaburi-dam in Laos has been pursued justified by the imperative for Laos (as its government points out) to safeguard export income and to bring economic growth to the poverty stricken country with few alternative incomes. Moreover, based on the case of Thailand, Mirumachi discusses this in terms of the prevailing logic of the 'hydraulic mission', that doesn't easily give way to reflexive modes of management. She also shows how different paradigms can exist simultaneously and how different institutions, such as the MRC and the Greater Mekong Subregion (GMS), are used in different roles to pursue sometimes incompatible objectives.

In his contribution Magee provides a concrete example of securitization of water management when he discusses the Chinese logic for hydropower expansion in the upper Lancang-Mekong. The Lancang cascade is part of China's Western Development Campaign which according to Magee represents an effort to remedy economic disparities between eastern and western China that run the risk of leading to widespread unrest. The issue is thereby internally securitized as economic development becomes a matter of national security and rhetorically of the survival of the state. Transboundary cooperation tends to remain within this logic and may even be reinforced when cooperation takes the shape of joint megaprojects, weather through bilateral or multilateral agreements (Chaps. 5 and 9).

Despite its focus on IWRM, the MRC cooperative arrangements seem to privilege large-scale development and its intergovernmental character has allowed national interests to predominate. Yet the problem is not just the incompatibility between national development imperatives such as the competing interventions of hydropower expansion in the upstream countries and fisheries in the downstream, as Keskinen et al. clarify, but how national interests are reduced to 'state interests' in terms of large-scale infrastructure for rapid economic growth without consideration of alternative voices promoting a more sustainable and participatory approach, an argument also Hirsch subscribes to. The Chinese case is even more extreme since its production of hydropower for electricity is argued to be of global interest due to the threat of climate change and the imperative to cut China's rapidly increasing carbon emissions (cf. Chap. 9 Magee).

The tendency to reduce conflicting interests to the level of the state may be further strengthened in a transboundary context. The question then is if this is accentuated by the particular cooperative framework, if it necessarily has to be so, or if it is actually possible to imagine it differently. That is if it is possible to imagine a cooperative framework that enables a political discussion about IWRM based on participation, in this particular regional context. As it seems, as above, issues become securitized rendering the state dominant or, as below, depoliticized concealing the issues of importance, either way creating processes obscuring the issues that would need to be laid bare and confronted.

10.1.3   IWRM Hiding Politics – The Art of Doing Business as Usual

As argued in the introduction and showed in some of the contributions to this book, focusing on the contradiction between IWRM and transboundary water management may hide dilemmas of water management within states, challenges that might to some extent resemble each other in the different countries (see Chaps. 1, 2, 5 and 9 this volume). The volume argues that it is imperative to make politics visible if we want to provide an opportunity for constructive debate on the futures of water resource use and management, not least if the grassroots shall have any chance to engage. The openness of the IWRM concept requires a focus on the politics involved in its continuous negotiation – openness can neither be fostered partially nor unevenly.

While regional water governance is important, and is typically not given sufficient attention, there may also be situations where the regional serves as an excuse for politicians (and other stakeholders) to avoid dealing with controversial local/national politics of water (cf. Chap. 6, Keskinen et al. this volume). Hence the pressure for regional perspectives may serve to place and bury issues at regional level, where inactivity may be a predictable outcome. Hence, in transboundary basins with explicit and institutionalised attempts at joint or coordinated development, the massive push for regional cooperation from the aid- and policy-community may – in a possibly counter-productive way – serve certain interests, and divert the issue of inadequate national and/or local water governance.

As such, the task of the MRC becomes – at times – a structurally impossible one, it is left to deal with the issues that cannot be worked out, which no state wants politically discussed, and possibly not even solved. The only way for MRC is then the technical approach, hoping it will pull politics along, but with limited political clout. Historically this has been the case, but as Öjendal and Mørck Jensen argue, the Xayaburi story – in which belatedly the MRC was instrumental in putting the process on the table and giving it its rightful political exposition – may provide a watershed (as Keskinen et al. also argue, but for other reasons) for the MRC to engage, or to trigger its members to engage, in core regional politics.

An additional dilemma, touched upon above, is that there is a tendency to reduce national interests to state interests, which leads to the exclusion of other voices *within countries* (cf. Chap. 6 Keskinen et al. this volume). It does not suffice to say that there is no such thing as national interest, but it is also important to unpack the notion of *state interest* as something homogenous. In this volume Magee shows how there are competing interests within the state by pointing at the competing interests of various ministries as well as China's quasi-governmental energy development corporations. Mirumachi discusses the Thai hydraucracy and shows how various parts of the Thai state pursue different agendas. Vietnam has experienced a similar internal process in relation to its role as upstream (vis-à-vis Cambodia) and downstream country (vis-à-vis the rest), where the energy department had different agendas as compared to the agriculture department. The MRC has been good at supplying technical information and knowledge as a base for better basin management. It has not been so strong in political monitoring, even less in policing free-riding (where each country may seek circumventing various parts of the MRC-framework).

As we saw above, one way of trying to desecuritize water resources management in order to move forwards with integration and sustainable development interventions has been to privilege cooperation on technical issues rather than political ones. This 'neo-functional' approach to river basin (and indeed regional cooperation) is a well-known, and not always very efficient *modus operandi* of overcoming politics that cannot be successfully negotiated directly. In fact, this may feed into the sovereignty logic, allowing business as usual by hiding contentious issues such as dam construction, rather than critically discuss them politically and provide possibilities for going beyond national/state interests.

IWRM holds a promise of broad and inclusive development and the exposition and promotion of the interests of different actors and sectors. As such it relies on participation and sharing of knowledge about the water system, its use and impacts thereof. IWRM provides a rhetoric that is ready-made to use, makes a lot of common sense, and provides a certain degree of good things for everybody. If we are content with that, rhetoric will be allowed to replace substance and where outsiders, non-experts, and citizens are robbed of access to the real process, which instead of being open for discussion will arrive as a given fact at a late stage, creating marginalization and discontent (cf. Chap. 8 Hirsch this volume). In order for IWRM to be a progressive tool in a transboundary context, this volume argues that we need to overcome the trap of de-politicization.

10.1.4   The Politics of Dialogue – Participation as Increasingly Pursued and Structurally Complex

The demand for dialogue and participation that comes with IWRM is potentially powerful. In one way or the other, all chapters in this volume point at the need for improved communication and dialogue, while most also show how participation has been marginalised (or not improved sufficiently) and hollowed out for issues treated in the transboundary context. Many questions as to the potential and nature of participation in this context remain, as do the challenges for a productive dialogue.

In his chapter, Hirsch discusses the centralizing and non-inclusive tendency of the MRC, but also sheds light on how counter-forces have opened up for a more participative approach. In 2008 the MRC renewed its commitment to public participation (cf. Chap. 4 Cooper this volume), as a result of development of the institution over time, a response both to donor pressure, new management, and an increasingly vocal civil society. Yet the possibility for such a process is questionable. The potential for institutions not designed to address the problems of the population as they perceive it, may not be well placed to attempt to do that. The somewhat detached character of the MRC is – in addition to that it is placed on regional level for regional concerns – a result of its heavy donor dependence as Hirsch tells us (cf. Chap. 8), but also due to their preference to work with a neo-functional approach.

It deserves to be observed that there are serious attempts made to 'ground' the activities of the MRC. For instance, Hirsch brings up the creation of sub-area committees, created as a way to remedy the absence of the MRC in the locales where people live, but points out that their composition is still heavily government-based and they are somewhat artificially created. This illuminates the dilemma of creating participation from above – possibly this should not be crafted from a regional body level. The establishment of RBOs (or the like) on sub-basin level as a form of bio-regionalism that relies on natural rather than administrative boundaries is another way of approaching the problem. This is established practice in many other contexts, however, the sub-basin RBOs in the Mekong are largely supported by bilateral or multilateral donors, thus largely consultant-designed and driven, rather than a result of stakeholders' definition of the problems. Again, crafting participation has its limits.

Furthermore, IWRM, Hirsch argues, has both centripetal and centrifugal tendencies. It has totalizing tendencies as it aims at geographical and sectorial integration requiring a holistic approach, which stands in contrast to its participatory focus on incorporating different stakeholders in an inclusive way. There are problems involved with mobilizing social engagement by local communities for a holistic approach that is difficult to make concrete and relevant in the local context. This indicates the importance of paying attention to the difference between various stakeholders as well as the need to create fora that have the potential to enable constructive dialogue. Hirsch shows that the most effective activities in terms of local community mobilization have been single-issue campaigns, primarily focused on opposing large-scale dam projects or other key issues rather than promoting IWRM perspectives. Bottom-up approaches by NGOs, although they may not have a participatory focus or make use of the IWRM discourse, are by Hirsch considered to be integral to a stakeholder-oriented approach.

Although Hirsch and Cooper further show how the MRC's basin development plan process has been opened up to a wider range of groups, the politics of participation keep posing important challenges. For instance, who is a relevant stakeholder? Under which condition are stakeholders allowed to participate?

Obviously there are big differences in terms of opportunities to participate in open dialogue for different stakeholders, both between and within countries (cf. Chap. 4). Certain types of stakeholders such as international NGOs are privileged, while others, such as local communities may face structural impediments.

Although all stakeholders are invited to the negotiation table there are other ways of limiting debate, e.g. Hirsch exemplifies how non-desired debates are excluded from the BDP-forum by influential actors (cf. Chaps. 3 and 8). Moreover, the structure with highly condensed technical information presented without prior possibility to digest the material constitutes yet another exclusionary mechanisms, where the "space for meaningful debate is already closed", in the words of the MRC itself (in Chap. 8).

If we recognize, as Hirsch does, that "[t]he real world of social engagement is one of negotiated outcomes, not of contrived platforms for orchestrated decision making", it becomes necessary to find novel foras and mechanisms for inclusion of local communities as stakeholders in the debate. In search for this Lazarus et al. explore E-flows as a tool for such a dialogue, to "facilitate participatory negotiation", to bring together various stakeholders to discuss sharing of benefits, costs and risks.

So, the regional cooperation and its institutions display severe shortcomings in terms of creating an enabling environment for participation in the deeper sense of the word. Having said that, it should be beyond reproach that MRC has made an effort to trigger dialogue and participation the last years. However, there is severe distrust between key stakeholders in the system. For instance, Cooper notes that some riparians see MRC (and the surrounding discourse backed up by donors and NGOs) as dominated by 'anti-dam' sentiments. The totally opposite view has simultaneously been aired, where the MRC is seen as the *de facto* executor of the pre-determined mainstream scenario, where dams and other large-scale modernizing interventions are predictable. We would agree with Cooper when she argues for the role of the MRC as facilitator when it comes to creating trust between stakeholders, and with Öjendal and Mørck Jensen that something is achieved by the efforts by MRC at participation and dialogue, imperfect as it may be.

However, although most would agree that participation and dialogue is far more advanced in the region as compared to what it used to be, the above also shows that participation is structurally difficult to pursue in a meaningful way in a contested transboundary setting. We may have to rethink how 'voice' can be supported and strengthened in cases like this. Rather than crafting participation from above, we need to foster stakeholder participation from below, to benefit from the local perspective, based on engagement and allowing the local dynamic to have impact.

10.1.5 Knowledge Management as a Key to Make a Difference – and Its Inevitable Politicisation

Sharing of technical and ecological knowledge and information is often considered a productive way to start a dialogue since it is assumed to be less politically tainted, hence less controversial. However, such assumptions may obfuscate the political character of knowledge, its production, its dissemination as well as its interpretation. Also 'simple' facts such as water flows carry political implications. More explicitly, impact assessments (cf. Chap. 3) become politicized under a guise of 'scientific-ness', indeed, at times it goes as far as to replace politics, policymakers ending up arguing whether or not impact assessments are correct or not.

Keskinen et al. point at the relative inconsistency of different assessments of the same scenario due to different assumptions used and differences in models and tools (also reflecting the complexity of the issue at hand). Many of the assessments have a narrow focus and "tend to 'compartmentalize' the environment and social systems into selected indicators and sectors" (Chap. 6). Despite these problems, the assessments are used as a basis for development planning, for example the estimation by the World Bank and the Asian Development Bank that there is "considerable potential for development of the Mekong water resources" (WB and ADB 2006). While "current assessment procedures tend to overly 'scientize' and depoliticize the knowledge production" (cf. Chap. 6 Keskinen et al.), the political nature of knowledge (and possibly the weakness of 'science') is obfuscated, as is the fact that they usually provide important room for interpretation for political and economic interests.

Moreover, IWRM is perceived as problematic in its focus on a "common, relatively pre-defined management approach for different scales and contexts" (cf. Chap. 6). Lacking transparency and reliability (and how 'reliable' can impact assessments actually be when there always is an interest and when matters studied are this complex?) turn assessments susceptible to misuse and political capture. To avoid technical solutions concealing politics behind knowledge production and particular interpretations, Keskinen et al. emphasize the need for parallel processes and methods for management and impact assessments. A proliferation of methods, they argue, may bring divergences to the fore, and at the same time avoid unrealistic estimates and increase credibility and transparency. The E-flows concept (Chap. 7) may be an approach doing exactly that, fruitfully complementing (and contesting) other studies with other methodologies. An acceptance of different assessment models rather than a reduction and homogenization could imply a re-politicization of knowledge production, bringing its political and moral nature to the fore (Chap. 6). To ensure that mediating tools such as E-flows "become an integral part of river basin management" by creating a space for dialogue where costs and benefits for different sectors are made explicit is a worthwhile effort (cf. Chap. 7).

As Lazarus et al. discuss E-flows as knowledge-based tools for negotiating water management, on the basis of a number of examples in the Mekong Region, they also display how 'threatening' new (not controllable) knowledge generation devices are, and how explicit knowledge generation also project political power. Correctly, they point out that there is a wide range of ways to interpret and approach the issue of E-flows, and that the process is likely to be contentious and involve trade-offs. However, the purpose of E-flows as a knowledge generation method is to make explicit the costs and benefits for different sectors so that informed negotiations can take place. Despite the problems involved in pursuance of E-flows analysis in the region (with little history in sustainable development) several lessons can be drawn already at this stage. So far, in the Mekong basin, the effects of E-flows analysis have been limited not least due to reluctance to openly display results, which again goes to show how 'dangerous' knowledge is when there is a lack of consensus among the riparian countries on how to interpret the outcome. Lazarus et al. argue for the importance of translating and internalizing tools such as E-flows to allow them to make a difference.

10.2   Political Imperatives and the Eternal Call for Holism

The contributions to this volume have in different ways addressed the dilemma of pursuing IWRM in a transboundary context. Together they help us unpack the perceived incompatibility between an integrative participatory approach and international dynamics relying on a sovereignty logic in a contested space. In a more forward-looking manner this volume also tries to address ways to work with the dilemma by paying attention to politics. The overall political regime in which the proposed basin management is to take place, is paramount. Regional water governance does not live in a political vacuum, and at some point, what can be done, is not what donors want or rely on the ideologies on water management that are in fashion, but what regional politics allow for. Having said that, politics is possible to alter; in fact, while constituting the limits of the possible, there is a tangible development of both regional politics and views on basin management. Below follows a brief recap of the key themes emerging in the unpacking of the IWRM-Transboundary dilemma, pulled from the sub-themes above.

In isolation RBOs will neither solve the dilemma of basin governance, nor be able to introduce IWRM on a basin level without a corresponding political will on national and local level. The opposite to over-belief in the RBOs, would be the capture of the development dynamic by the state, arguing the prerogative of economic growth and nationally based modernization. This is a pattern in the Mekong basin, although it is typically pursued below the MRC-radar, avoiding confrontation and occasionally even insight. Interestingly, and providing a historical 'watershed' opportunity, the Xayaburi-dam controversy falls in line with the historical pattern of a national prerogative, but deviates from the habit of eluding open MRC-treatment.

While a decade of IWRM-policies have laid the groundwork which made possible the relative open Xayaburi process, and must be seen as a success, IWRM may more often than not provide a terminology where controversial politics can be imbedded, made invisible, and partially disappear in good-sounding rhetoric. This is a threat to proper basin management, and a way – possibly an unconscious one – to obfuscate transparent management which also reduces the ability for the non-expert to participate in debates. MRC has attempted to remedy its obvious historical deficit of participatory approaches, but with its, after all, limited approach, it has not managed to address the structurally determined dilemmas involved in enabling a public 'voice' to be heard.

Finally, judging from the above, there is a great game going on, being played by most stakeholders in the region. In different roles, various stakeholders seek to visualize and hide perspectives respectively; some seek to place debates where they can be controlled and others aim to bring the same debates into the spotlight for a more open discussion. Obviously, knowledge that is not politically controlled becomes controversial: tightly controlled impact assessments carried out by actors receptive to deliver according to a certain agenda is common, whereas, for instance, the acceptance of an E-flow analysis is kept at arms-lengths distance. Either way, knowledge is politicized and turned into politics in itself. This is also a reason why a volume like this serves a purpose beyond its academic value.

At the end of the day, IWRM is not bad, basin cooperation is necessary, and the transboundary issues will not go away any time soon. While our scrutiny here is a critical reflection on how power and visions (politics and development) merge (or not), interact, and appear mutually exploitative, this is not an argument for not continue working with these aspects of water management. To the contrary, water governance is likely to grow more contested and even more central to key stake-holders in the decade to come. That is why we need the appropriate tools to pursue it openly and efficiently, and to study it critically and creatively. Possibly we should listen to Keskinen et al. asking for a 'pause' to strengthen the development dialogue, hence being able to produce more sustainable solutions.

References

Earle A, Jägerskog A, Öjendal J (eds) (2010) Transboundary water management principles and practice. Earthscan, London

Mirumachi N (2010) Study of conflict and cooperation in international transboundary river basins: the twins framework. PhD thesis, LSE, London

WB/ADB (June 2006) Joint working paper on future directions for water resources management in the Mekong river basin, Mimeo

Wolf AT (1997) 'Water wars' and water reality: conflict and cooperation along international waterways. In: Lonergan SC (ed) Environmental change, adaption, and security. Kluwer, Boston, pp 251–265

Index

J. Öjendal et al. (eds.), *Politics and Development in a Transboundary Watershed:*
The Case of the Lower Mekong Basin, DOI 10.1007/978-94-007-0476-3,
© Springer Science+Business Media B.V. 2012

Printed by Books on Demand, Germany